Raphael Aardoom

**Monopnictogen Ferrocenes as Ligands for Asymmetric Catalysis**

Raphael Aardoom

# Monopnictogen Ferrocenes as Ligands for Asymmetric Catalysis

Südwestdeutscher Verlag für Hochschulschriften

**Impressum / Imprint**

Bibliografische Information der Deutschen Nationalbibliothek: Die Deutsche Nationalbibliothek verzeichnet diese Publikation in der Deutschen Nationalbibliografie; detaillierte bibliografische Daten sind im Internet über http://dnb.d-nb.de abrufbar.

Alle in diesem Buch genannten Marken und Produktnamen unterliegen warenzeichen-, marken- oder patentrechtlichem Schutz bzw. sind Warenzeichen oder eingetragene Warenzeichen der jeweiligen Inhaber. Die Wiedergabe von Marken, Produktnamen, Gebrauchsnamen, Handelsnamen, Warenbezeichnungen u.s.w. in diesem Werk berechtigt auch ohne besondere Kennzeichnung nicht zu der Annahme, dass solche Namen im Sinne der Warenzeichen- und Markenschutzgesetzgebung als frei zu betrachten wären und daher von jedermann benutzt werden dürften.

Bibliographic information published by the Deutsche Nationalbibliothek: The Deutsche Nationalbibliothek lists this publication in the Deutsche Nationalbibliografie; detailed bibliographic data are available in the Internet at http://dnb.d-nb.de.

Any brand names and product names mentioned in this book are subject to trademark, brand or patent protection and are trademarks or registered trademarks of their respective holders. The use of brand names, product names, common names, trade names, product descriptions etc. even without a particular marking in this works is in no way to be construed to mean that such names may be regarded as unrestricted in respect of trademark and brand protection legislation and could thus be used by anyone.

Coverbild / Cover image: www.ingimage.com

Verlag / Publisher:
Südwestdeutscher Verlag für Hochschulschriften
ist ein Imprint der / is a trademark of
AV Akademikerverlag GmbH & Co. KG
Heinrich-Böcking-Str. 6-8, 66121 Saarbrücken, Deutschland / Germany
Email: info@svh-verlag.de

Herstellung: siehe letzte Seite /
Printed at: see last page
**ISBN: 978-3-8381-3628-8**

Zugl. / Approved by: Zürich, ETH, Diss., 2012

Copyright © 2013 AV Akademikerverlag GmbH & Co. KG
Alle Rechte vorbehalten. / All rights reserved. Saarbrücken 2013

# Contents

    Abstract . . . . . . . . . . . . . . . . . . . . . . . . . . . . . iii
    Zusammenfassung . . . . . . . . . . . . . . . . . . . . . . . . vi
    Introductory Remarks . . . . . . . . . . . . . . . . . . . . . . ix
    Abbreviations . . . . . . . . . . . . . . . . . . . . . . . . . . . xi

**1 Introduction**                                                       **1**
   1.1   Ferrocenes . . . . . . . . . . . . . . . . . . . . . . . . . . 1
   1.2   Transfer Hydrogenation . . . . . . . . . . . . . . . . . . . 5
   1.3   Aim of this Thesis . . . . . . . . . . . . . . . . . . . . . . 10

**2 Aminoferrocenes**                                    **11**
   2.1   Introduction . . . . . . . . . . . . . . . . . . . . . . . . . 11
   2.2   Project Idea and Synthetic Approach . . . . . . . . . . . 17
   2.3   Results . . . . . . . . . . . . . . . . . . . . . . . . . . . . 19
   2.4   $\beta$-Diketiminato Ligands . . . . . . . . . . . . . . . . . . . 32
   2.5   Ferrocenyl Ureas . . . . . . . . . . . . . . . . . . . . . . . 48
   2.6   Additional Investigations . . . . . . . . . . . . . . . . . . 55
   2.7   Conclusions . . . . . . . . . . . . . . . . . . . . . . . . . 57

**3 Phosphinoferrocenes**                              **59**
   3.1   Introduction . . . . . . . . . . . . . . . . . . . . . . . . . 59
   3.2   Ligand Synthesis . . . . . . . . . . . . . . . . . . . . . . . 63
   3.3   Gold(I)-Complexes in Asymmetric Catalysis . . . . . . . 68
   3.4   Ferrocenyl-Tethered $Ru^{II}$-Complexes for Asymmetric Catalysis . . . . . . . . . . . . . . . . . . . . . . . . . . . 85
   3.5   Additional Investigations . . . . . . . . . . . . . . . . . . 99

|   |     | 3.6 Conclusions | 105 |
|---|-----|-----------------|-----|
| **4** | **General Conclusions and Outlook** | | **107** |
|   | 4.1 | Aminoferrocenes | 109 |
|   | 4.2 | Phosphinoferrocenes | 110 |
|   | 4.3 | Outlook | 111 |
| **5** | **Experimental Part** | | **112** |
|   | 5.1 | General Remarks | 112 |
|   | 5.2 | Syntheses | 117 |
|   | 5.3 | Catalyses | 151 |
| **6** | **Appendix** | | **xiii** |
|   | 6.1 | Crystallographic Data | xiv |
|   | **Bibliography** | | **xxxvi** |

# Abstract

The first part of this thesis describes the synthesis and reactivity of a chiral aminoferrocene. The main steps for preparing a sterically demanding primary aminoferrocene starting from Ugi's amine include substitution of its amino group, carboxylation of the ferrocene and subsequent Curtius degradation. After five synthetic steps, the aminoferrocens is obtained in 67% overall yield.

Further functionalization of the amine is best carried out *via* condensation reactions with carbonyl compounds. For example the condensation of two equivalents of amine with acetylacetone yielded a bis(ferrocenyl)-$\beta$-diketimine; this class of compounds was unprecedented at the time of its synthesis. Probably due to the high steric demand of the ferrocenyl substituents, attempts to prepare corresponding complexes were unsuccessful.

A range of urea derivatives were prepared by quenching the isocyanate formed after Curtius-degradation with water or amines. Preliminary experiments showed promising activities and enantioselectivities in indium(0)-mediated asymmetric allylations of hydrazones.

[Scheme: PhCH=N-NHBz + allyl bromide, with 10 mol-% ferrocenyl urea catalyst (bearing Ph-CH(Me)- group and 2-MeO-C6H4-NH-C(O)-NH- group), 1.7 eq In⁰, toluene, −10 °C, 24 h → PhCH(NHNHBz)CH2CH=CH2, 44%, er 68:32]

In the second part of this thesis, the chiral ferrocenyl backbone was used to prepare a small library of monophosphino ferrocenes starting from the corresponding halides.

[Scheme: Ferrocenyl-CH(Ar)-Br → (n-BuLi, R2PCl, −78 °C to reflux) → Ferrocenyl-CH(Ar)-PR2]

R = Ph, Ar = Ph (67%)
R = Ph, Ar = 3,5-xylyl (70%)
R = Ph, Ar = 3,5-di-*tert*-butylphenyl (68%)
R = Ph, Ar = 2,4,6-mesityl (58%)
R = Ph, Ar = 1-naphthyl (70%)
R = Cy, Ar = Ph (80%)
R = *i*-Pr, Ar = Ph (76%)

[Further transformations: with AuCl(SMe2) → Ferrocenyl-CH(Ar)-PPh2-Au-Cl; Ar = Ph (quant), 1-naphthyl (quant).
With [RuCl2(p-cymene)]2, toluene, reflux, 15 h → Ferrocenyl-tethered Ru complex]

R = Ph, Ar = Ph (95%)
R = Ph, Ar = 3,5-xylyl (85%)
R = Cy, Ar = Ph (80%)
R = *i*-Pr, Ar = Ph (85%)

These ligands were coordinated to gold(I) and ruthenium(II) precursors. No catalytic applications for the linear gold(I) compounds could be found, *inter alia* due to unexpected difficulties in the halide abstraction step. Upon coordination to ruthenium(II), ferrocenyl-tethered complexes were obtained. When applying these compounds as catalysts in asymmetric transfer hydrogenation, moderate activities and enantioselectivities were obtained. The catalyst performance was strongly dependent on the used base and counter ions.

$$\text{Ph}\overset{\text{O}}{\underset{}{\bigwedge}} \xrightarrow[\substack{\text{NH}i\text{-Pr}_2,\ i\text{-PrOH} \\ 80\ °\text{C, 20 h}}]{\substack{5\ \text{mol-\%} \ [\text{Ru catalyst}] \\ 5\ \text{mol-\%} \ \text{Et}_3\text{OSbF}_6}} \text{Ph}\overset{\text{OH}}{\underset{}{\bigwedge}} \quad \substack{59\% \\ \text{er } 72{:}28}$$

# Zusammenfassung

Im ersten Teil dieser Dissertation wird die Synthese eines chiralen Aminoferrocens sowie seine Reaktivität behandelt. Ausgehend von Ugi's Amin wurde durch Substitution der Aminogruppe sowie Carboxylierung des Ferrocens und anschliessendem Curtius-Abbau ein sterisch anspruchsvolles, primäres Aminoferrocen hergestellt. In fünf präparativen Schritten wurde das Amin in einer Gesamtausbeute von 67% erhalten.

Kondensationsreaktionen des Amins mit Carbonylverbindungen stellten sich als bevorzugte Methoden zur Funktionalisierung heraus. So wurde beispielsweise durch Kondensation mit Acetylaceton ein Bis(ferrocenyl)-$\beta$-diketimin erhalten, eine zum Zeitpunkt der Herstellung noch unbekannte Substanzklasse. Versuche zur Herstellung von Koordinationsverbindungen schlugen möglicherweise aufgrund der sterisch anspruchsvollen Substituenten fehl.

Durch eine Modifikation des Curtius-Abbaus wurden ferrocenylsubstituierte Harnstoffderivate erhalten. Erste Untersuchungen zeigten vielversprechende Aktivitäten und Selektivitäten in der Indium(0)-vermittelten asymmetrischen Allylierung von Hydrazonen.

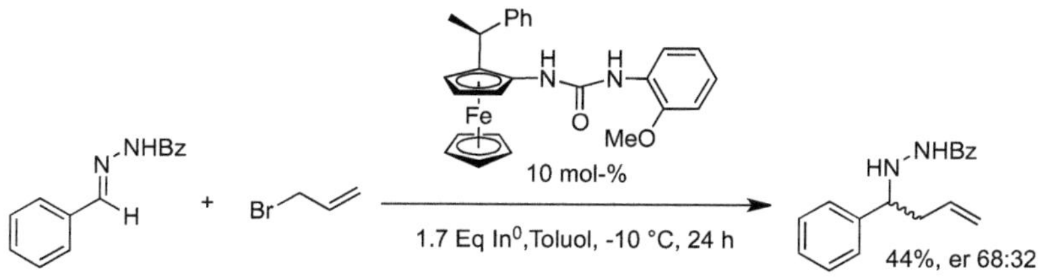
Ph
H
H
N
N
Fe
O
MeO
N
NHBz
H
+
Br
10 mol-%
1.7 Eq In⁰,Toluol, -10 °C, 24 h
HN
NHBz
44%, er 68:32

Im zweiten Teil dieser Dissertation wurde das asymmetrische Rückgrat des Aminoferrocens beibehalten. Ausgehend von Ferrocenylhalogeniden wurden eine Reihe Monophosphinoferrocene hergestellt.

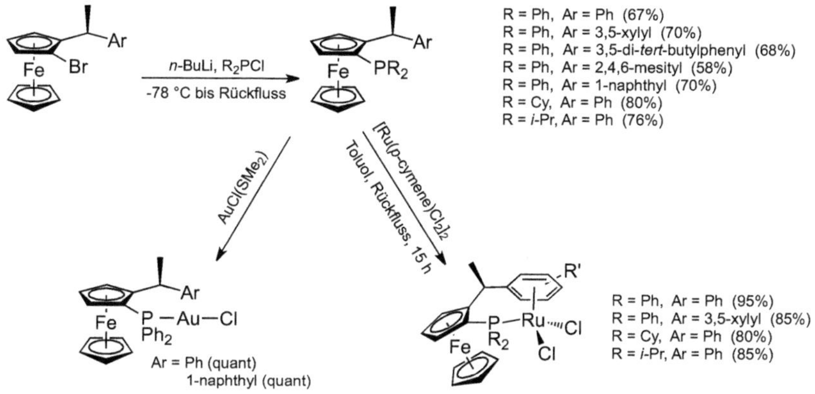
Ar
Fe
Br
n-BuLi, R2PCl
-78 °C bis Rückfluss
Ar
Fe
PR2
R = Ph, Ar = Ph (67%)
R = Ph, Ar = 3,5-xylyl (70%)
R = Ph, Ar = 3,5-di-tert-butylphenyl (68%)
R = Ph, Ar = 2,4,6-mesityl (58%)
R = Ph, Ar = 1-naphthyl (70%)
R = Cy, Ar = Ph (80%)
R = i-Pr, Ar = Ph (76%)
AuCl(SMe2)
[Ru(p-cymene)Cl2]2
Toluol, Rückfluss, 15 h
Ar
Fe
P
Au
Cl
Ph2
Ar = Ph (quant)
1-naphthyl (quant)
R'
P
Ru
Cl
R2
Cl
Fe
R = Ph, Ar = Ph (95%)
R = Ph, Ar = 3,5-xylyl (85%)
R = Cy, Ar = Ph (80%)
R = i-Pr, Ar = Ph (85%)

Die Liganden konnten unter anderem an Gold(I)- und Ruthenium(II)-Vorläufer koordiniert werden. Während für die linearen Gold(I)-Verbindungen keine katalytische Anwendung gefunden werden konnte (*inter alia* aufgrund der heiklen Chloridabstraktion), bilden die Liganden mit Ruthenium(II) *Tether*-Komplexe, welche in der asymmetrischen Transferhydrierung moderate Aktivitäten und Selektivitäten erreichten. Die Leistungen der Katalysatoren weisen hohe Basen- und Gegenionenabhängigkeiten auf.

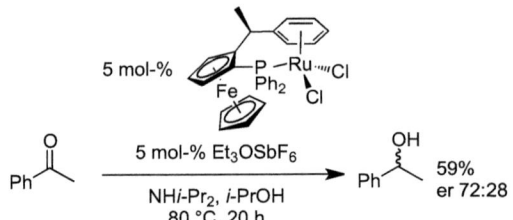

# Introductory Remarks

This thesis was submitted to ETH Zurich in 2012 as Diss. ETH No. 20635 and was accepted on the recommendation of Prof. Dr. Antonio Togni and Prof. Dr. Hansjörg Grützmacher.

The original document is available at the ETH E-Collection via http://e-collection.library.ethz.ch

## Structure of this Thesis

While this dissertation picks up where the author's master's thesis ended,[1] it explores a variety of related research topics before reaching some of its initial goals. As diverging as the chosen paths may seem, the "leitmotiv", i. e. the chiral ferrocenyl backbone, is present throughout. Furthermore, the structure of the thesis largely corresponds to the chronological course of action, which makes less than obvious transitions from one topic to another more comprehensible. It is organized as follows:

CHAPTER ONE introduces the reader briefly to ferrocene and its derivatives. Stereochemical definitions and synthetic pathways to chiral ligands are discussed, as well as their application to asymmetric catalysis. Transfer hydrogenation is discussed and the initial goals of the thesis are stated.

CHAPTER TWO treats aminoferrocenes and the synthetic approach towards the envisioned target ligand, and unexpected derivatives, such as ureas and $\beta$-ketimines which eventually led the author to consider the synthesis of phosphinoferrocenes.

CHAPTER THREE concerns monophosphinoferrocenes, their gold(I) and ruthenium(II) complexes and—in conclusion—shows their applications to asymmetric transfer hydrogenation.

In order to minimize leafing back and forth, topic-specific introductions are given throughout all chapters.

# Stereochemistry and Atom Numbering

Although both (R)- and (S)-Ugi's amine served as starting materials for the synthetic work, reaction schemes show only (R)-derivatives. ORTEPIII-representations of crystal structures show the mesured enantiomers. When directly comparing opposite enantiomers, e. g. in tables or for structural overlays, structures and structural data of (S)-derivatives were inverted.

Similarly, catalyses were carried out with both enantiomers. The enantiomeric ratios obtained with (S)-derivatives were inverted.

For the sake of comparison and to prevent overly crowded illustrations, the atom numbering in crystal structures follows the pattern of the ligand presented below. Atoms deviating from this template are labeled individually.

# Abbreviations

| | |
|---|---|
| acac | acetylacetonato |
| Ar | aryl group |
| ATR | attenuated total reflection |
| $BAr_F^-$ | tetra(3,5-bis(trifluoromethyl)phenyl)borate |
| BINAP | 2,2'-bis(diphenylphosphino)-1,1'-binaphthalene |
| BINOL | 1,1'-bi-2-naphthol |
| Bn | benzyl |
| Bu | butyl |
| Cbz | carboxybenzyl |
| cod | 1,5-cyclooctadiene |
| Cp | cyclopentadiene |
| Cp* | 1,2,3,4,5-pentamethylcyclopentadiene |
| d | day(s) |
| dipp | 2,6-di-*iso*-propylphenyl |
| DMF | *N*,*N*-dimethylformamide |
| DMSO | dimethylsulfoxide |
| DPPA | diphenylphosphoryl azide |
| dppf | 1,1'-bis(diphenylphosphino)ferrocene |
| dppp | 1,3-bis(diphenylphosphino)propane |
| dr | diastereomeric ratio |
| ee | enantiomeric excess |
| EI | electron ionization |
| en | 1,2-diaminoethane |
| eq | equivalent |
| er | enantiomeric ratio |
| Et | ethyl |
| etb | ethylbenzoate |
| Fc | ferrocenyl |
| FC | flash chromatography |
| FID | flame ionization detector |
| h | hour |

| | |
|---|---|
| HiRes | high resolution |
| HNacNac | pentane-2,4-diimine |
| Hz | Hertz |
| IR | infrared |
| $J$ | coupling constant |
| MALDI | matrix-assisted laser desorption / ionization |
| Me | methyl |
| MeOH | methanol |
| min | minute(s) |
| mmb | methyl-(2-methyl)benzoate |
| MS | molecular sieves |
| NMR | nuclear magnetic resonance |
| Np | naphthyl |
| Ph | phenyl |
| ppfa | $(S_p)$-1-diphenylphosphino-2-[$(R)$-(1-$N,N$-dimethylamino)ethyl]ferrocene |
| ppm | parts per million |
| $p$-TsOH | $p$-toluenesulfonic acid |
| PyBOX | pyridyl bis(oxazoline) |
| rac | racemic |
| r. t. | room temperature |
| TBS | *tert*-butyldimethylsilyl |
| THF | tetrahydrofuran |
| tht | tetrahydrothiophen |
| TLC | thin layer chromatography |
| TMEDA | $N,N,N',N'$-tetramethyl-1,2-diaminoethane |
| TMSNHOTMS | $N,O$-bis(trimethylsilyl)hydroxylamine |
| Ts | *para*-toluenesulfonyl; *tosyl* |
| TsDPEN | $N$-(*para*-toluenesulfonyl)-1,2-diphenyl-1,2-diaminoethane |

# Chapter 1

# Introduction

This chapter gives an overview about ferrocenes in general, as well as their use as ligands in asymmetric catalysis. The reader is introduced to transfer hydrogenation, bifunctional ruthenium(II)-$\eta^6$-arene complexes and tethered catalysts.

## 1.1 Ferrocenes

Since its serendipitous discovery in the middle of the last century,[2, 3] ferrocene and its derivatives have become indispensable compounds throughout chemistry and materials science.[4–7] While the term *ferrocene* was deduced from *benzene* and chosen to point out the aromatic character of the molecule,[8] its three dimensional structure[9–12] accounts for the success of its derivatives as ligands in asymmetric catalysis.[6, 13, 14]

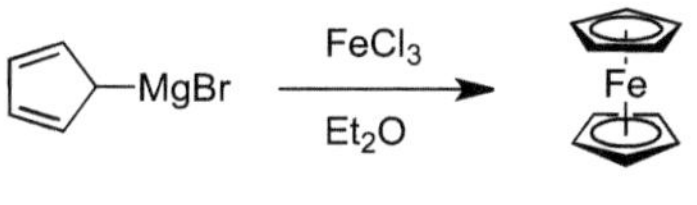

**Scheme 1:** Synthesis of ferrocene as reported in 1951 by Kealy and Pauson.[2]

In contrast to planar aromatic systems, introducing two different substituents on the same ferrocene ring system automatically yields chiral molecules. This novel concept of asymmetry, *planar chirality*, was soon extensively explored.[15, 16]

**Stereochemistry in Ferrocenes**

As planar chirality represented a concept that had not been treated in the classic sequence rule introduced by Cahn, Ingold and Prelog,[17–19] Schlögl proposed a simple rule for the assignment of the absolute configuration of ferrocenes:[20]

*The observer looks along the principal axis of the molecule so that, in the case of ferrocene derivatives, the more highly substituted ring is directed towards him, whereby the priority of the groups is decisive. The substituents are then, as usual, arranged in decreasing order of priority according to the sequence rule. If more than three groups are present, only the three with highest priority are considered; the choice of symbol (R) or (S) depends on the resulting direction (clockwise or counterclockwise).*[20]

This rule has been accepted as a standard for planar-chiral metallocenes. As modern derivatives of ferrocenes often contain several stereogenic units (central, planar and phosphorus chirality), subscripts are used to indicate the type of chirality, i. e. p for planar and P for phosphorus chirality. These conventions are presented in Scheme 2.

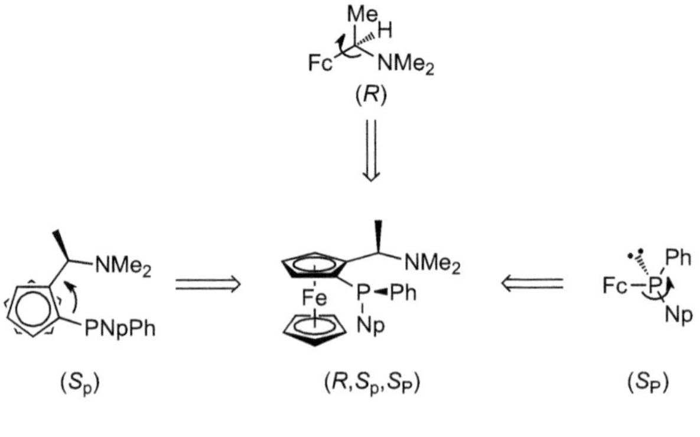

**Scheme 2:** Assignment of the absolute configuration in ferrocenes.[20]

## 1.1.1 Diastereoselective *ortho*-Functionalization of Ferrocenes

As its aromatic character allows electrophilic substitution reactions, various derivatives of ferrocene were reported shortly after its discovery.[8, 15, 16, 21] Although diastereoselective preparation of 1,2-disubstituted ferrocenes *via* the directing effect of chiral additives was known before,[20, 22–24] Ugi's simple method for preparing enantiomerically pure *N,N*-dimethylaminoethylferrocene (Ugi's amine **1**) was the first to allow the synthesis of enantiomerically pure planar-chiral ferrocenes in an efficient manner on a large scale.[25–27]

As shown in Scheme 3, the amine on the side chain acts as a directing group when lithiating Ugi's amine (**1**). Using this principle for diastereoselective *ortho*-lithiation of ferrocenes, other chiral directing groups were developed during the early 1990s, such as sulfoxides,[28] acetals[29] or oxazolines[30] (Scheme 3).

Although alternatives are known, the field of asymmetric ferrocenyl ligands is still dominated by the derivatives of Ugi's amine. While one may attribute this to a twenty year head start, the amine has other advantages. Besides its availability, it is also quite inert towards racemization, and many of its derivatives are air-stable and can be purified chromatographically. Another very important feature is the possibility to substitute the amino functionality by nucleophiles. In the case of the unsubstituted amine, this occurs *via* an $S_N1$ mechanism with retention of configuration, in which the intermediate carbocation is stabilized by the iron center (Scheme 4).[31, 32] Recent results show that this finding is not neccessarily true for substitutions of other α-dimethylamino ferrocenes; depending on the substituents, both inversion and retention of configuration may take place.[33]

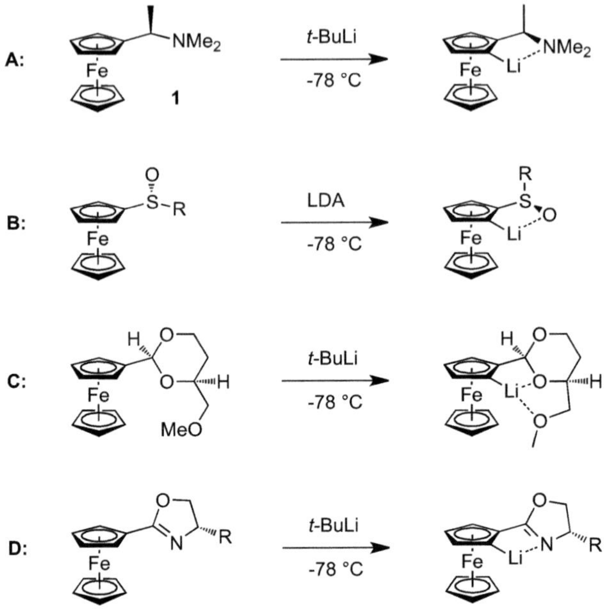

**Scheme 3:** Diastereoselective *ortho*-lithiation of ferrocenes using chiral directing groups. A: Ugi's amine **1**;[25, 26] B: sulfoxides;[28] C: acetals;[29] D: oxazolines.[30]

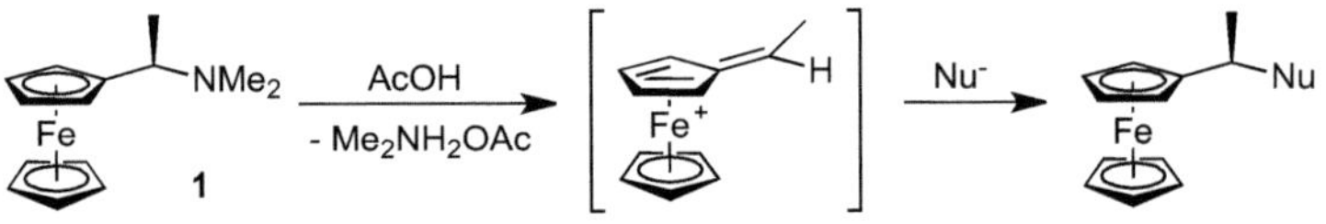

**Scheme 4:** Nucleophilic substitution in acetic acid occurs with retention of configuration.

Many efficient ferrocenyl ligand classes are based on Ugi's amine. The most renowned one, *Josiphos*, can be prepared in just two steps as presented above: after introduction of one phosphine group *via* diastereoselective *ortho*-lithiation to give *ppfa*, the amino group is substituted by dicyclohexylphosphine to give the final ligand (Scheme 5).[34]

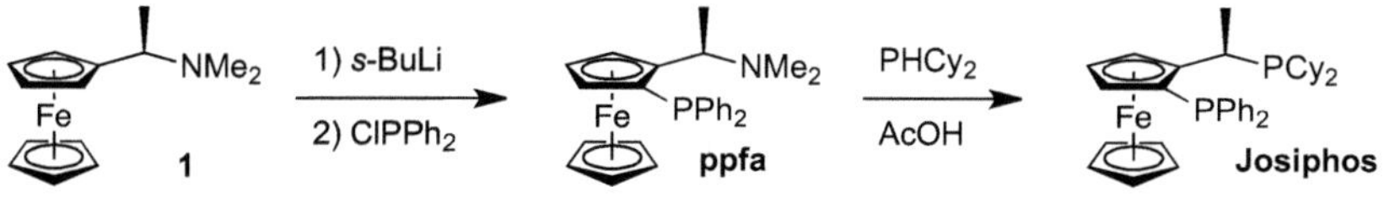

**Scheme 5:** Synthesis of Josiphos according to Togni *et al.* [34]

## 1.2 Transfer Hydrogenation

The reduction of unsaturated compounds *via* formal addition of dihydrogen ranks among the most important reactions in synthetic chemistry. While molecular hydrogen of course represents the cleanest source, it is also highly flammable and reactions often require the use of high-pressure equipment. In contrast to these limitations, the use of organic hydrogen donors allows the reactions to be carried out conveniently in standard glassware at mild temperatures. Transfer hydrogenation was first introduced in the mid-nineteen twenties; the aluminum *iso*-propoxide catalyzed reduction of ketones by alcohols under basic conditions is currently known as the Meerwein-Ponndorf-Verley reaction (Scheme 6).[35–37]

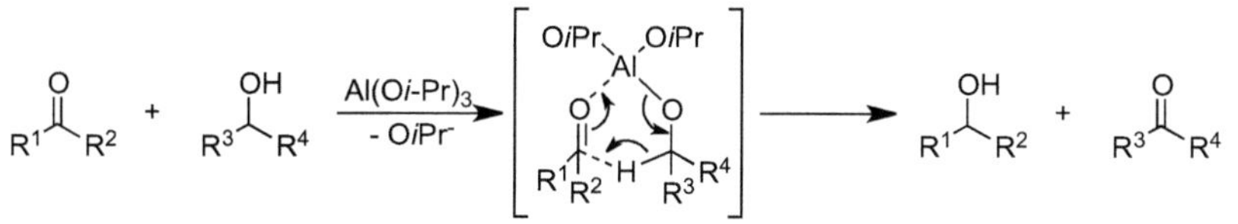

**Scheme 6:** The Meerwein-Ponndorf-Verley reaction.[35–37]

As shown in Scheme 6, the mechanism of this reduction is a simple one. Under basic conditions, the carbonyl compound and an alcoholate coordinate to a Lewis acid and a hydride transfer occurs. The same mechanism may be formulated for the selective dehydrogenation of alcohols, which is referred to as an Oppenauer oxidation.[38]

Commonly used hydrogen donors are *iso*-propanol, formic acid (or formate salts) and the azeotropic mixture of formic acid and triethylamine. While the alcohol is a fairly safe and non-toxic compound, it has has one major drawback. As its oxidation product, acetone, remains in the reaction mixture, it becomes a hydrogen acceptor itself. Depending on the thermodynamics of the system, a large excess of alcohol may be needed in order to reach high conversions.* Alternatively, the nascent acetone may be distilled off constantly. Formic acid, on the other hand, is oxidized to carbon dioxide, which immediately leaves the system and thereby allows full reduction of the carbonyl compound. However, its range of application is reduced due to its incompatibility with most of the catalysts used. Less common hydrogen donors include primary alcohols, e. g. ethanol or butanediol, which can be oxidized to the corresponding esters[40–44] and lactones.[41, 42] Also, Hantzsch esters have been used as reductants.[45] An advantage of such alternative hydrogen sources is that the overall reaction is practically irreversible. A selection of hydrogen sources and their corresponding products are shown in Scheme 7.

### 1.2.1 Asymmetric Transfer Hydrogenation with Bifunctional Ruthenium(II)-Arene Catalysts

The efficient preparation of chiral compounds is a fundamental requisite throughout the pharmaceutical, agrochemical, fragrance and food industries. During the last decades, asymmetric transfer hydrogenation has become one of the most popular tools for the generation of new stereocenters and has therefore been reviewed extensively.[46–50] Consequently, this introduction does not intend to cover this vast field

---

*The reduction of acetophenone using fifty equivalents of *iso*-propanol reaches an equilibrium at 96% conversion.[39]

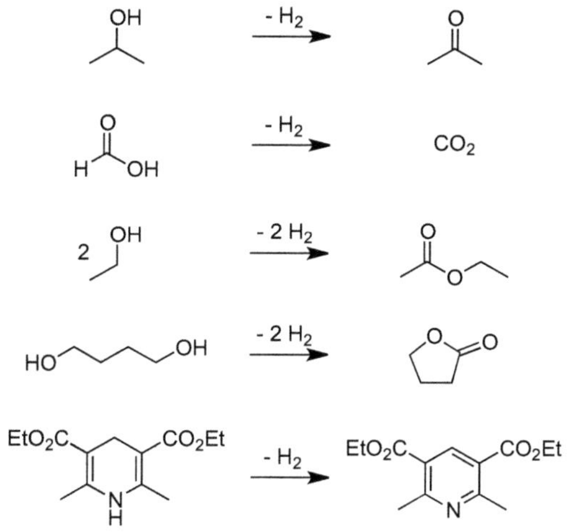

**Scheme 7:** Hydrogen sources and their corresponding products.

comprehensively but gives the background needed for this thesis.

The first metal-catalyzed asymmetric hydrogen transfer was reported in 1950, when Doering and Young published an asymmetric version of the Meerwein-Ponndorf-Verley reduction.[51] Using chiral alcohols as hydrogen donors, up to 22% ee was obtained (Scheme 8).

Throughout the years, only a handful of successful catalyst systems were developed until Noyori et al. [52] presented his bifunctional ruthenium(II) catalyst **2** in 1995 (Figure 1). Due to its outstanding performance, it is still regarded as the benchmark for asymmetric transfer hydrogenation.

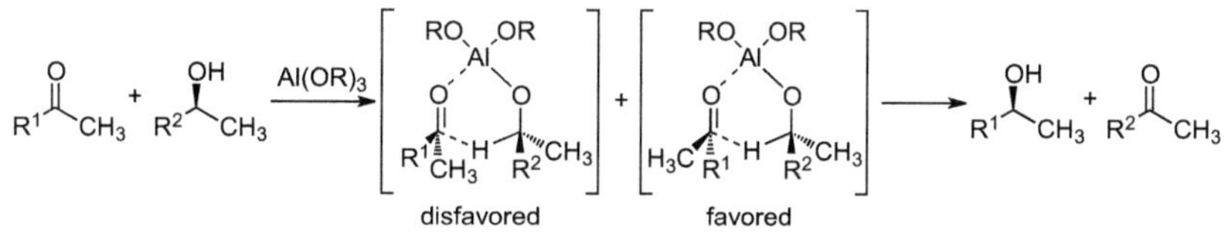

**Scheme 8:** First asymmetric transfer hydrogenation as reported by Doering and Young.[51]

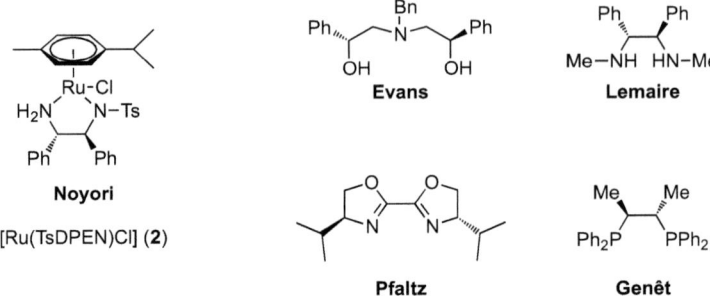

**Figure 1:** Noyori's catalyst **2**,[52] and successful pre-Noyori ligands.[53–56]

The most important feature of Noyori's catalyst is the presence of an amino group bonded to the metal. As shown in Scheme 9, this allows for a metal-ligand bifunctional reaction mechanism in which both the metal and nitrogen atoms actively take part in the hydrogenation step. Although the substrate does not coordinate directly to the metal, the six-membered cyclic transition state is well defined and accounts for excellent stereoselectivities.[57]

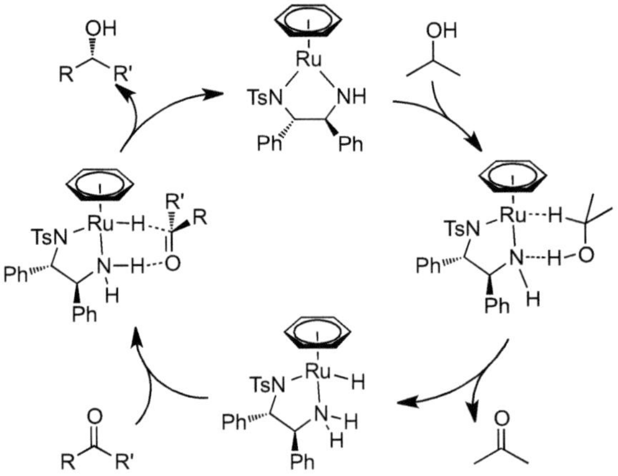

**Scheme 9:** Metal-ligand bifunctional reaction mechanism according to Noyori.[57]

Inspired by the huge success of this new paradigm, several ligands were designed specifically to enable bifunctional catalysis (Figure 2).

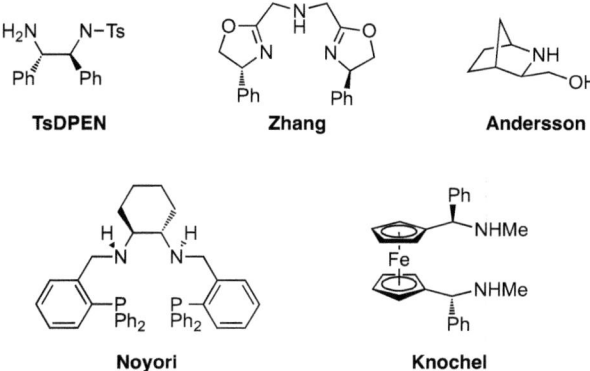

**Figure 2:** Representative ligands designed to imitate the functionality of TsDPEN.[58–61]

Among these newly developed catalysts, the tethered analogues developed by Wills[62, 63] are amongst the few examples that are actually able to compete with Noyori's archetype **2**. The tether provides a more rigid catalyst structure, thereby improving its stereoselectivity and the stability of the complex. Thus, both the tethered complex **3** and its 'reverse-tethered' analogue **4** display faster reaction rates and higher enantioselectivity than its non-tethered analogues (Figure 3).

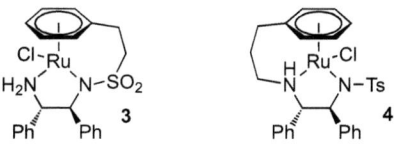

**Figure 3:** Wills' tethered and 'reverse-tethered' catalysts **3** and **4** outperform their non-tethered analogues.[62, 63]

## 1.3 Aim of this Thesis

The aim of this thesis is to combine the advantages of bifunctional catalysts (i. e. **2**) and the enhanced performance of Wills' tethered analogue **4** with the asymmetric induction of planar-chiral ferrocenes (Scheme 10). A ferrocenyl-tethered complex would have a sterically demanding, chiral group in close proximity to the active site, thus the incorporation of a chiral diamine might not be required in order to achieve high selectivities in asymmetric transfer hydrogenation.

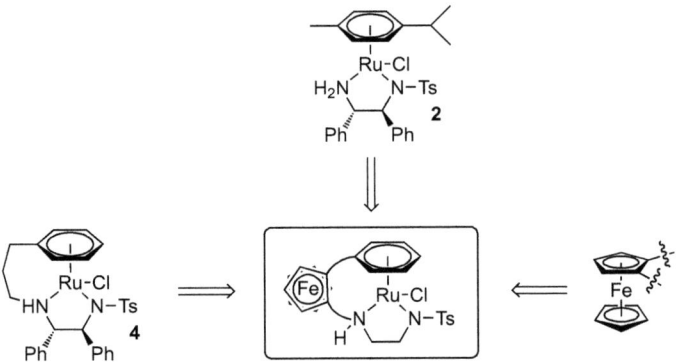

**Scheme 10:** General stuctural motive of the target complex.

# Chapter 2

# Aminoferrocenes

The straight-forward pathway towards the envisioned complex, the reactivity of the obtained primary aminoferrocene and its limitations are discussed. Sections 2.4 and 2.5 focus on the syntheses and applications of derivatives of the aminoferrocene: bis(ferrocenyl)-$\beta$-diketimine and ferrocenyl ureas.

During the author's master's thesis, the unoptimized synthesis of the primary aminoferrocene was established and preliminary experiments concerning its reactivity were performed.[1] Part of the synthetic work concerning ferrocenyl ureas was carried out by Katrin Niedermann and Jolanda Winkler.

## 2.1 Introduction

Aminoferrocene was first synthesized in 1955 by Nesmeyanov *et al.* [64] As shown in Scheme 11, monolithiated ferrocene was reacted with *O*-benzylhydroxylamine to give the primary amine in 25% yield.

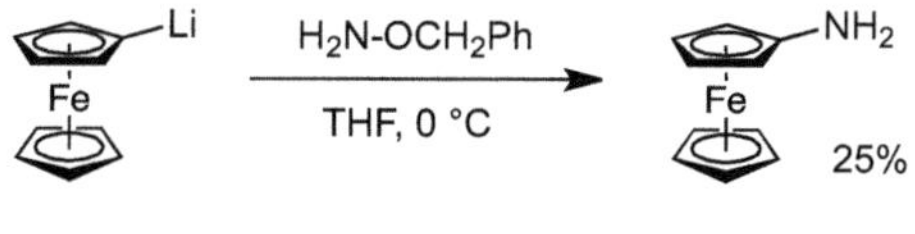

**Scheme 11:** Preparation of aminoferrocene as published by Nesmejanov *et al.* [64]

This preparation remained the only one-pot route to aminoferrocene until Hessen and co-workers published another procedure nearly five decades later (Scheme 12).[65] In the meantime, various alternative syntheses had been described, mostly based on copper-mediated coupling reactions or *via* Curtius-degradation, as shown in Schemes 13 and 14, respectively.

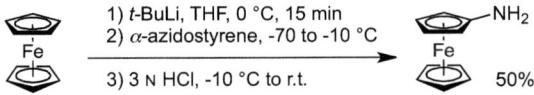

**Scheme 12:** Preparation of aminoferrocene as published by Hessen and co-workers.[65]

Copper-mediated preparation of aminoferrocene was reported early on, again by Nesmeyanov *et al.* Ferrocenylboronic acid was reacted with cupric phthalimide to give *N*-ferrocenylphthalimide, which was reduced with hydrazine to give the primary amine in 30% yield over two steps.[66] In an alternative approach by Arnold and co-workers, cuprous iodide was used for the preparation of 1,1'-diazidoferrocene from 1,1'-dibromoferrocene and sodium azide. Palladium/charcoal catalyzed reduction yielded the corresponding diamine in an overall yield of 45%.[67]

The synthesis of aminoferrocene starting from ferrocenylcarboxylic acid *via* Curtius degradation was reported by Arimoto and Haven shortly after the first report on aminoferrocene (Scheme 14).[21] Just like for the other procedures discussed above, the overall yield was low; 20% of the final product was obtained. This may partially be explained by it being a four-step procedure including comparatively harsh conditions.

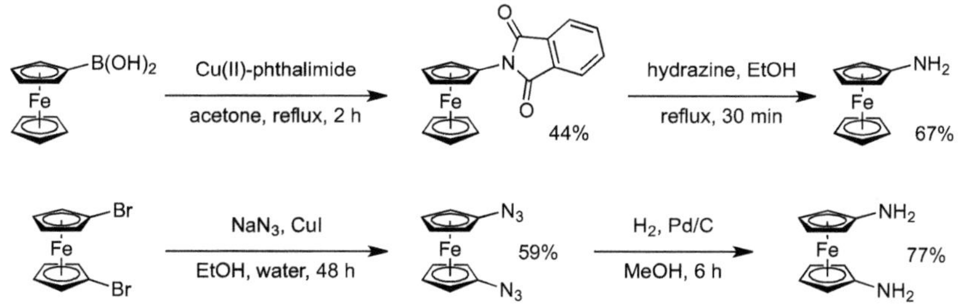

**Scheme 13:** Copper-mediated preparations of aminoferrocenes as published by Nesmeyanov et al. [66] (top) and Arnold and co-workers [67] (bottom).

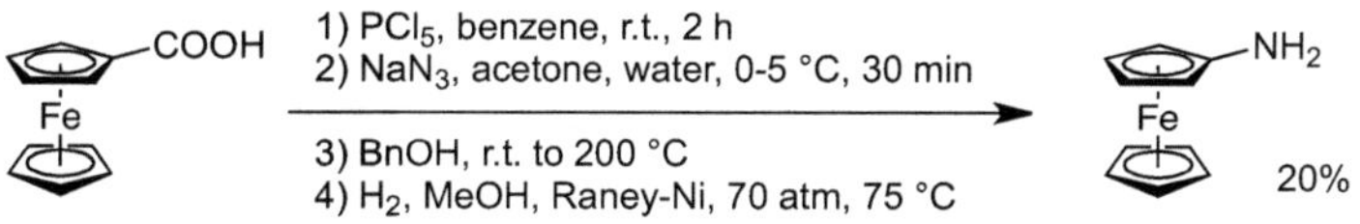

**Scheme 14:** Preparation of aminoferrocene via Curtius degradation as published by Arimoto and Haven.[21]

## 2.1.1 Synthesis of Chiral Aminoferrocenes

Many of the above mentioned methods have been used for the preparation of enantiomerically pure aminoferrocenes. Independent of the applied concept of stereodiscrimination, the synthetic pathways usually include two steps to introduce the amino group; additional steps are then needed to remove the directing group or to introduce other functional groups.

Selected synthetic strategies are shown in Schemes 15 and 16. As it was already described for the non-chiral aminoferrocenes, yields are mostly mediocre. A positive exception is Richards' synthesis via a nitroferrocene. When neglecting the loss in yield due to the conversion of the oxazoline directing group to an ester, the amino group is introduced in 86% yield.[68]

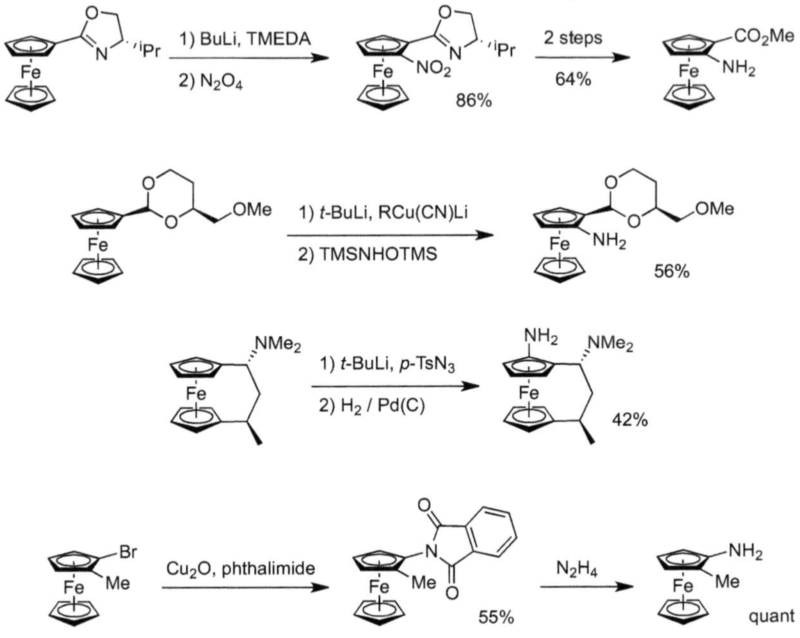

**Scheme 15:** Syntheses of chiral aminoferrocenes *via* direct Cp-N bond formation and subsequent reduction as published by Richards,[68] Kagan,[69] Erker,[70] and Barybin[71] (top to bottom).

A novel approach was published recently by Metallinos *et al.* [72] Instead of having a chiral directing group to allow for stereoselective *ortho*-amination, copper-mediated coupling of L-proline hydantoin to iodoferrocene introduces the amine as part of a chiral directing group (Scheme 16).

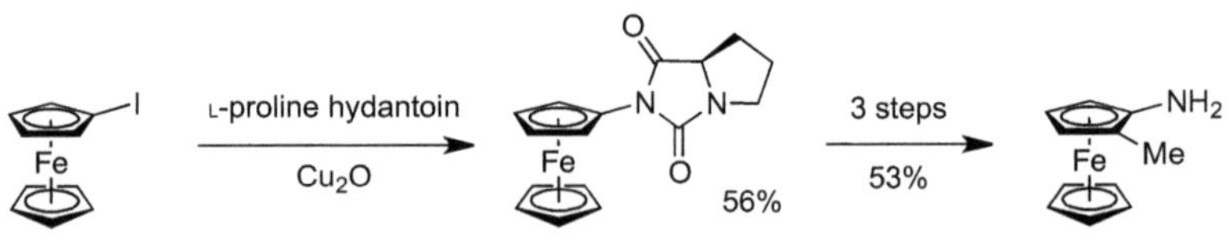

**Scheme 16:** Synthesis of chiral aminoferrocenes starting from a non-chiral precursor as published by Metallinos *et al.* [72]

Chiral aminoferrocenes were also prepared *via* Curtius degradation by Bertogg *et al.* [73] (Scheme 17). After following Kagan's route for stereoselective *ortho*-functionalization, the chiral acetal was converted to the primary amine in six steps. An alternative approach starting from Ugi's amine leads *via* 1-carboxyl-2-vinyl-ferrocene to the carbamate-protected aminoferrocene. The vinyl group may be functionalized later on.

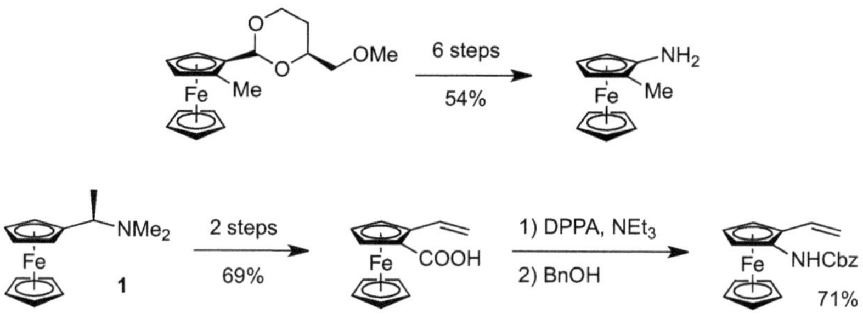

**Scheme 17:** Preparation of chiral aminoferrocenes *via* Curtius degradation as published by Bertogg *et al.* [73, 74]

## 2.1.2 Complexes and Catalytic Applications of Aminoferrocenes

Although aminoferrocenes have been known for more than fifty years, only a handful of complexes and even less catalytic applications have been published. When searching the *SciFinder*, *reaxys* and *CSD* databases* for complexes of monodentate ferrocenyl amine, only three molybenum complexes were found in which aminoferrocene acts as a redox-active imido-ligand.[75] In all other cases, aminoferrocenes are parts of polydentate ligands[76] or the nitrogen coordinates as part of a functional group, such as *N*-heterocyclic carbenes[73] or an amidinato moiety.[74] Representative complexes are shown in Figure 4.

---
*July 21, 2012; CCDC ConQuest 4.14 search in CSD version 5.33 (Nov. 2011).

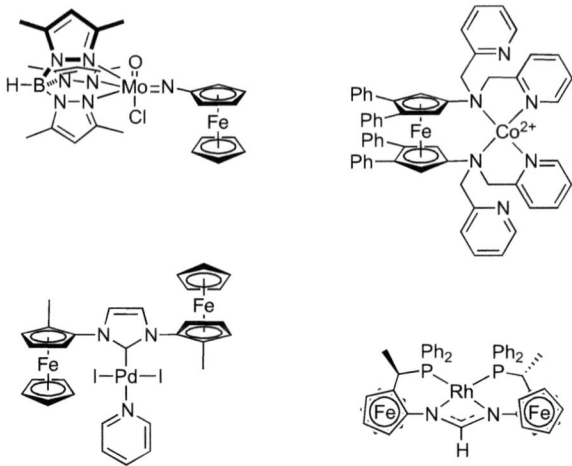

**Figure 4:** Representative complexes containing aminoferrocenes.[73–76]

Among the small number of complexes, even less catalytic applications have been found, e. g. Nazarov cyclization,[77] allylic alkylation,[77] intramolecular arylation[73] and olefin polymerization.[78] In many cases, the catalyst performance was low to moderate. One of the few examples in which high yields and selectivities were reached is shown below (Scheme 18).

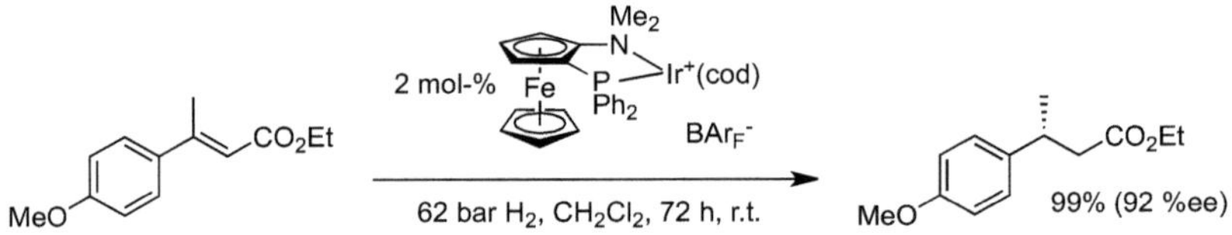

**Scheme 18:** Asymmetric hydrogenation of alkenes using a P-N ligand.[79]

## 2.2 Project Idea and Synthetic Approach

As discussed in Section 1.3, the initial goal of this thesis was the synthesis of a ferrocenyl-tethered ruthenium(II) complex inspired by Noyori's piano-stool complex [Ru(TsDPEN)Cl] (**2**)[52] and Wills' tethered analogue **3**.[63] Based on the available starting materials and the scientific background of our group, it was decided that the tentative synthetic route would use Ugi's amine (**1**) as starting material. Also, the synthesis would implement Bertogg's procedure for the synthesis of primary ferrocenylamines.[73, 74] To ensure a clean Curtius rearrangement, as few functional groups as possible should be present besides the carboxylic acid. A retrosynthetic analysis considering these guidelines shows that the target ligand **10** should be accessible in six steps, as depicted in Scheme 19.

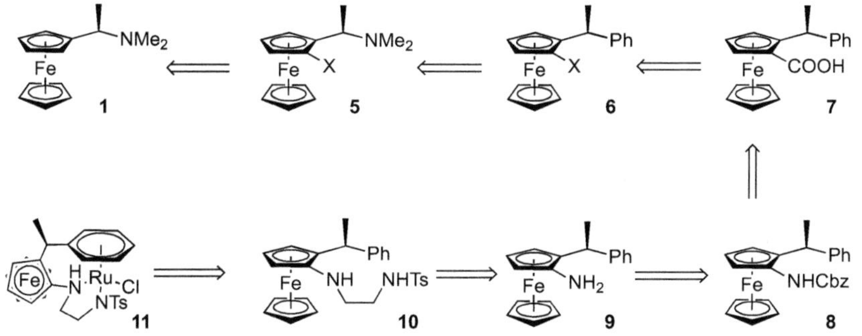

**Scheme 19:** Retrosynthetic analysis for the synthesis of target compound **11**.

The retrosynthetic analysis shows the high modularity of the ligand, manifested by two intermediates: the aryl-substituted halide **6** and aminoferrocene **9**. The aryl group might be introduced according to a protocol intended for the synthesis of Taniaphos-type ligands (Scheme 20).[33, 80] While Knochel and co-workers used 2-bromo-iodobenzene in order to produce bidentate ligands, only monohalide benzene derivatives were intended to be used. A wide range of such compounds is available and they should allow for both steric and electronic fine-tuning of the projected catalyst. The final step of the synthesis—functionalization of the primary amine—should open even more doors; a vast library of ligands was axpected to be accessible.

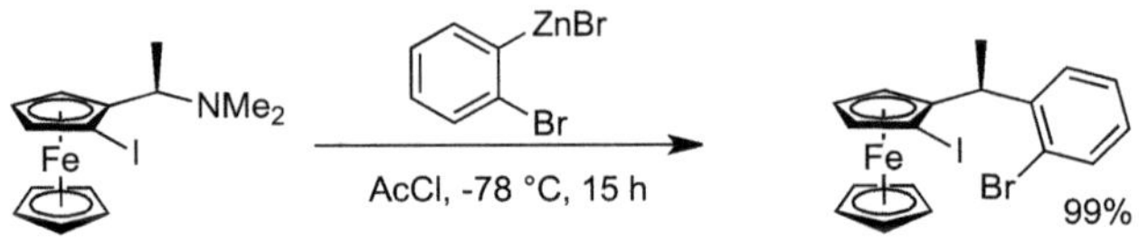

**Scheme 20:** Substitution of the amino function as published by Knochel and coworkers.[80]

While the afore mentioned steps account for the variety of ligand, they were not thought to be the most challenging. Knochel's protocol is well established and high yielding,[80] and even though the functionalization of aminoferrocenes had not been investigated in depth so far, a wide range of reactions, such as alkylation, condensation, or metal catalyzed cross-coupling reactions could be tested. The aminoferrocenes reported by Bertogg et al.,[73, 74] however, are quite sensitive compounds. Therefore, their synthesis and isolation were expected to be the key steps of the project.

## 2.3 Results

### 2.3.1 Synthesis of a Chiral Aminoferrocene

#### 2.3.1.1 Halogenation and Substitution of the Amino Moiety

For the halogenation of Ugi's amine (**1**), the known procedure[81–83] was further optimized. Instead of 1,2-dibromotetrafluoroethane, which is an ozone depletant and gives carcinogenic tetrafluoroethene as a side product during the reaction, less harmful 1,2-dibromotetrachloroethane was used as bromide source to give ferrocenylbromide **12**. As shown in Scheme 21, the iodo derivative **13** was prepared in similar fashion.

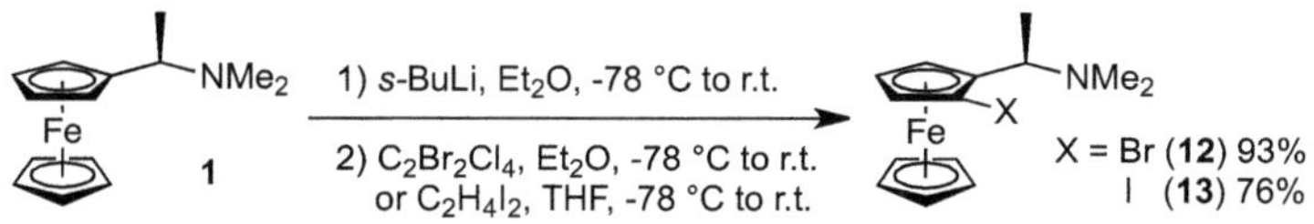

**Scheme 21:** Halogenation of Ugi's amine.

Introducing the aryl group on the side chain proved to be more difficult than expected. In contrast to the high yields reported by Knochel and co-workers,[80] synthetically useful yields could only be reached after tedious optimization of the protocol for the substitution of the dimethylamino group with phenylzinc bromide. Best results were obtained when working on at least millimolar scale; the reaction was carried out with up to 10 g of starting material. The arylzinc reagents as well as most of the parent Grignard reagents had to be prepared *in situ*, which may at least account partially for the varying yields. When commercially available diphenylzinc or a solution of phenylzincbromide was used, no product could be isolated. The *in situ* preparation of the reagents usually gave a very viscous slurry, therefore an oversized stirring bar had to be used. After an acceptable protocol was established for the substitution of the amino group of bromo-Ugi amine **12** to the corresponding bromide **14**, a range of derivatives were prepared. Besides the iodide **15**, which may be a good substrate for introducing substituents *via* cross-coupling reactions, six additional aryl groups were introduced. The yields are summarized in Table 2.1.

Table 2.1: Exchange of the amino group.

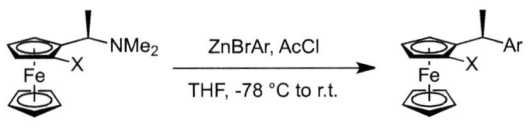

| entry | X | Ar | product | yield /[%] |
|---|---|---|---|---|
| 1 | Br | Ph | 14 | 81 |
| 2 | I | Ph | 15 | 62 |
| 3 | Br | 3,5-xylyl | 16 | 99 |
| 4 | Br | 2,4,6-mesityl | 17 | 68 |
| 5 | Br | 3,5-di-*tert*-butylphenyl | 18 | 71 |
| 6 | Br | 3,5-bis(trifluoromethyl)phenyl | 19 | 36 |
| 7 | Br | 1-naphthyl | 20 | 72 |
| 8 | Br | 8-fluoro-1-naphthyl | 21 | 47 |

**2.3.1.1.1 Crystal Structures** Except for iodide **15** and the 3,5-di-*tert*-butylphenyl substituted bromide **18** (which is an oil), the stuctures of the newly prepared compounds were determined by X-ray diffraction. ORTEPIII representations are given in Figure 5; bond and torsion angles are summarized in Table 2.2. As shown in the structural overlay (Figure 6), the structures are quite similar. Except for the steric repulsion between the *ortho*-methyl group of the mesityl substituent and the cyclopentadienyl unit in the case of compound **17**, the main structural differences appear to be caused by crystal packing effects.

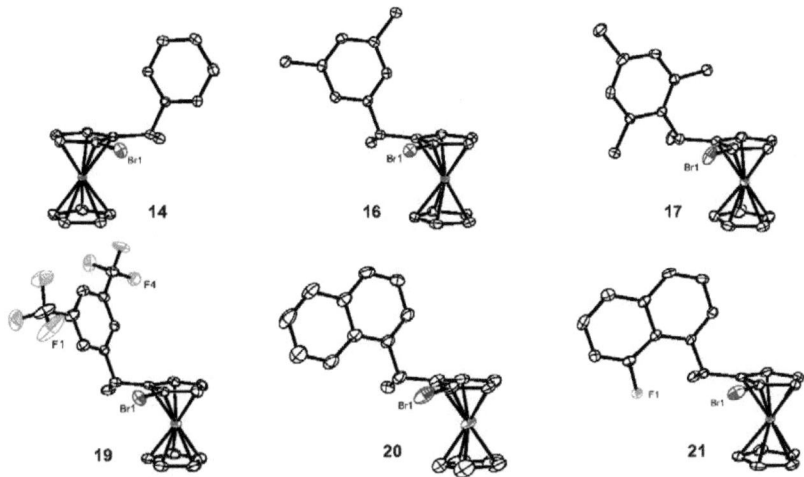

**Figure 5:** ORTEPIII representation of bromides **14**, **16**, **17**, **19**, **20** and **21**. Hydrogen atoms are omitted for clarity, thermal ellipsoids are set to 50% probability. Selected bond and torsion angles are given in Table 2.2.

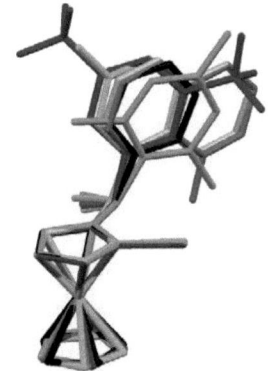

**Figure 6:** MERCURY-generated structural overlay of bromides **14**, **16**, **17**, **19**, **20** and **21**. Where neccessary, structures were inverted to show the ($S,S$)-derivatives.

Table 2.2: Selected bond and torsion angles of compounds **14, 16, 17, 19, 20** and **21**.[a]

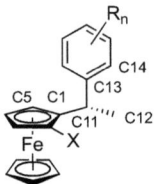

|  | 14 | 16 | 17[b] |
|---|---|---|---|
| C1–C11–C12 /[°] | 112.81(10) | 112.51(18) | 112.1(2) / 112.1(2) |
| C1–C11–C13 /[°] | 111.31(9) | 112.10(17) | 113.9(2) / 115.0(2) |
| C12–C11–C13 /[°] | 110.31(9) | 110.36(17) | 112.0(2) / 110.6(2) |
| C5–C1–C11–C12 /[°] | −20.25(16) | −28.4(3) | −14.1(4) / −10.4(4) |
| C5–C1–C11–C13 /[°] | 104.38(13) | 96.7(3) | 114.3(3) / 117.1(3) |
| C1–C11–C13–C14 /[°] | 141.03(11) | 145.46(19) | 133.4(2) / 136.4(2) |

|  | 19 | 20[c] | 21 |
|---|---|---|---|
| C1–C11–C12 /[°] | 113.3(3) | 113.0(4) | 112.1(3) |
| C1–C11–C13 /[°] | 109.4(3) | 110.5(5) | 110.6(2) |
| C12–C11–C13 /[°] | 111.9(3) | 111.2(4) | 111.2(2) |
| C5–C1–C11–C12 /[°] | −29.6(5) | −26.1(5) | −21.7(4) |
| C5–C1–C11–C13 /[°] | 96.1(4) | 99.2(6) | 103.0(3) |
| C1–C11–C13–C14 /[°] | 111.0(4) | 142.3(5) | 150.4(3) |

[a] Angles normalized for (S,S)-derivatives. [b] The asymmetric unit contains two molecules. [c] The substituted Cp-ring is heavily disordered.

## 2.3.1.2 Carboxylation, Curtius Degradation and Deprotection of the Amino Group

The synthetic procedures towards the primary ferrocenylamine are all based on the work of Bertogg *et al.* [74] Carboxylation of the halides **14** and **15** was achieved in quantitative yields when high-puritiy $CO_2$ was bubbled through a solution of the lithiated ferrocene to give the carboxylic acid **7** (Scheme 22). Yields dropped significantly when in-house dry ice or technical grade $CO_2$, purified by a Cu-catalyst and dried over $P_2O_5$, were used as $CO_2$ sources. The product was best purified by crystallization. Often, a displeasing odour emanated from the collected crystals, originating from valeric acid, which was formed during the reaction of excess $n$-butyllithium with carbon dioxide. This side product could be removed under high vacuum, leaving behind the carboxylic acid **7** as odorless orange crystals.

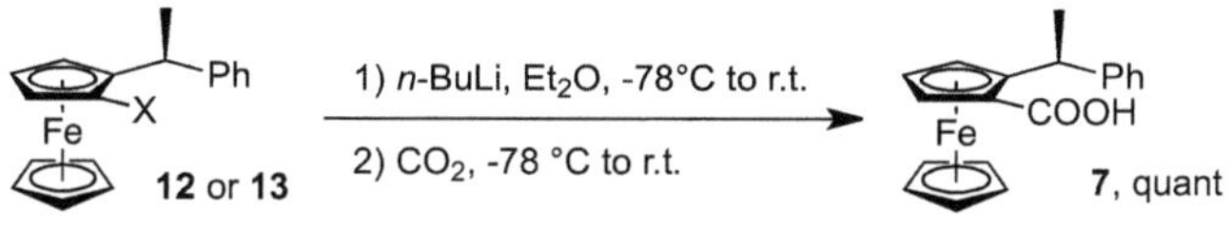

**Scheme 22:** Carboxylation of halides **14** and **15**.

Even though the reaction of the carboxylic acid **7** to the protected amine **8** *via* Curtius degradation can be carried out as a one-pot procedure, it is in fact a multi-step reaction. As shown in Scheme 23, the carboxylic acid is first converted into an acyl azide. Heating leads to the loss of nitrogen gas and the actual rearrangement to the isocyanate takes place. Quenching the reaction mixture with an alcohol yields a carbamate-protected amine.

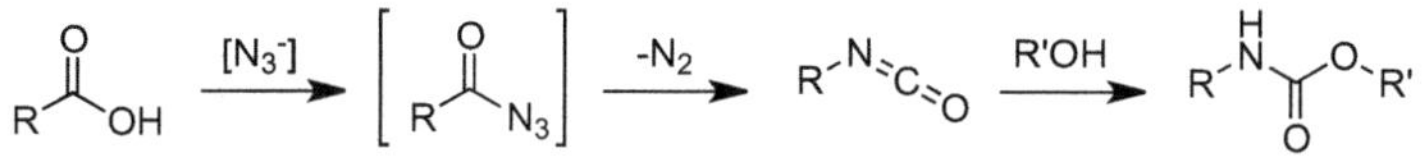

**Scheme 23:** General reaction scheme for the conversion of a carboxylic acid into a protected amine.

When putting theory into practice, monitoring the reaction by TLC turned out to be crucial. Especially the intermediate isocyanate proved to be rather sensitive to prolonged refluxing; delayed quenching led to a significant extent of decomposition.

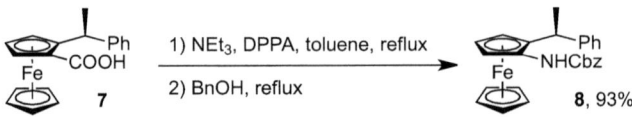

**Scheme 24:** Curtius degradation yielding carbamate **8**.

The amine was then deprotected under basic conditions (Scheme 25). In contrast to reports by Bertogg for vinyl or phosphine substituted aminoferrocenes,[77] the free amine **9** could be isolated in high yields and proved to be bench stable for years.

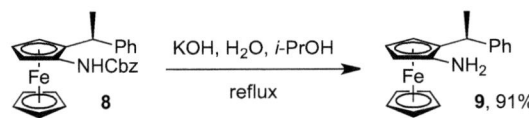

**Scheme 25:** Deprotection giving the air-stable ferrocenylamine **9**.

**2.3.1.2.1 Crystal Structures**  Carboxylic acid **7**, carbamate **8**, and the primary amine **9** could all be isolated in crystalline form, so it was decided to elucidate their solid-state structure by X-ray diffraction. The asymmetric units of both the carboxylic acid **7** (Figure 7) and the Cbz-protected amine **8** (Figure 8) each contain two symetrically independent molecules which interact with each other through hydrogen bonding. The primary amine **9**, however, shows no strong intermolecular interactions (Figure 9).

The oxo and hydroxy groups of the acids could easily be assigned by comparing the C–O bond lenghts (1.223(4) vs. 1.316(4) Å and 1.224(4) vs. 1.318(3) Å). Both acid groups are nearly coplanar with respect to the bound cyclopentadienyl units (torsions of 0.0(3)° and 3.5(3)°); the Cp-planes of the two molecules are at an angle of 7.73(3)° with respect to each other. At first glance, a $C_2$ axis might be recognized. However, this tentative symmetry is broken mainly by the conformation of the sidechains, especially the by torsion angles of the phenyl rings. (140.9(3) vs. 121.2(3)°). A molecular overlay is given in Appendix 6.1.8.

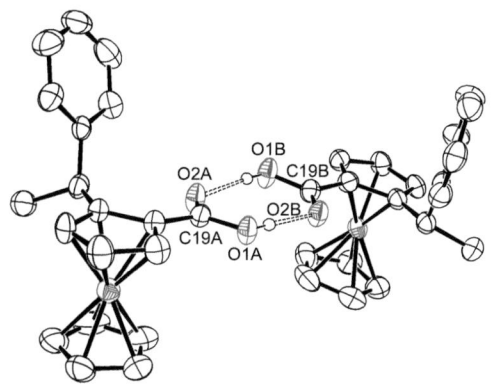

**Figure 7:** ORTEPIII representation of carboxylic acid **7**. If not part of hydrogen bonds, hydrogen atoms are omitted for clarity, thermal ellipsoids are set to 50% probability. Selected bond lengths [Å], bond and torsion angles [°]: O1A–C19A 1.316(4), O2A–C19A 1.223(4), C2A–C19A 1.470(4), O1B–C19B 1.318(3), O2B–C19B 1.224(4), C2B–C19B 1.462(4), O1A–O2B 2.621(4), O2A–O1B 2.646(4), O1A–C19A–O2A 123.4(3), C1A–C11A–C12A 112.1(3), C1A–C11A–C13A 110.1(2), C12A–C11A–C13A 110.5(2), O1B–C19B–O2B 122.7(3), C1B–C11B–C12B 111.6(3), C1B–C11B–C13B 113.5(3), C12B–C11B–C13B 109.2(3), C5A–C1A–C11A–C12A −21.9(4), C5A–C1A–C11A–C13A 101.6(3), C1A–C11A–C13A–C14A 140.9(3), O2A–C19A–C2A–C1A 3.7(5), C5B–C1B–C11B–C12B −28.1(5), C5B–C1B–C11B–C13B 95.8(4), C1B–C11B–C13B–C14B 121.2(3), O2B–C19B–C2B–C1B 0.2(5).

The structure of carbamate **8** does not convey many surprises either. As it could be expected, the substituents on the Cp-ring are in *exo* conformation. Two molecules interact *via* hydrogen bonding of the carbamate moieties, resulting in a tight package. As it was the case for the carboxylic acid, the two molecules do not exhibit large structural differences. A molecular overlay is given in Appendix 6.1.9.

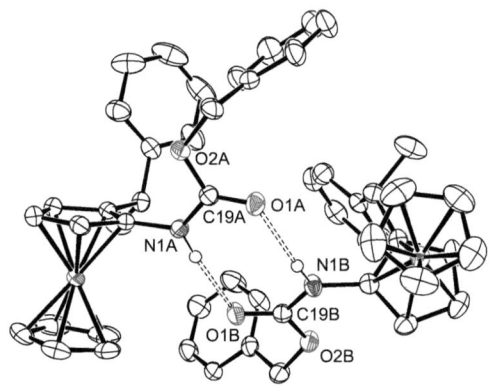

**Figure 8:** ORTEPIII representation of carbamate **8**. If not part of hydrogen bonds, hydrogen atoms are omitted for clarity, thermal ellipsoids are set to 50% probability. Selected bond lengths [Å], bond and torsion angles [°]: O1A–C19A 1.220(2), O2A–C19A 1.355(2), N1A–C2A 1.420(2), N1A–C19A 1.342(2), C1A–C11A–C12A 112.19(18), C1A–C11A–C13A 112.70(17), C12A–C11A–C13A 108.14(18), C2A–N1A–C19A 125.08(16), N1A–C19A–O1A 124.65(18), C5A–C1A–C11A–C12A −40.4(3), C5A–C1A–C11A–C13A 81.9(3), C1A–C11A–C13A–C14A 153.8(2), C1A–C2A–N1A–C19A 111.8(2), O1A–N1B 2.846(2), N1A–O1B 2.867(2), O1B–C19B 1.220(2), O2B–C19B 1.348(2), N1B–C2B 1.422(3), N1B–C19B 1.344(3), C1B–C11B–C12B 111.67(19), C1B–C11B–C13B 112.15(18), C12B–C11B–C13B 109.28(19), C2B–N1B–C19B 126.38(17), N1B–C19B–O1B 124.32(18), C5B–C1B–C11B–C12B −27.1(3), C5B–C1B–C11B–C13B 95.9(3), C1B–C11B–C13B–C14B 158.05(19), C1B–C2B–N1B–C19B 110.5(2)

Although hydrogen bonding has been reported for the crystal structures of aminoferrocene[84] and 1,1'-diaminoferrocene,[67] ferrocenylamine **9** shows no such interactions. Overall, its structure is quite similar to the structure of bromide **14**, differing mainly in the arrangement of the phenyl groups differ.

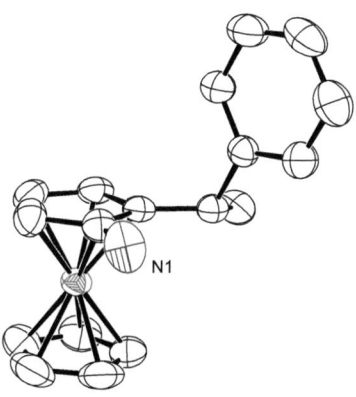

**Figure 9:** ORTEPIII representation of ferrocenylamine **9**. Hydrogen atoms are omitted for clarity, thermal ellipsoids are set to 50% probability. Selected bond lengths [Å], bond and torsion angles [°]: C2–N1 1.401(3), C1–C11–C12 113.26(16), C1–C11–C13 111.13(14), C12–C11–C13 110.60(15), C5–C1–C11–C12 −19.9(3), C5–C1–C11–C13 105.3(2), C1–C11–C13–C14 133.29(19).

## 2.3.2 Reactivity of the Amine

With the projected tridentate ligand in reach, it was first tried to alkylate the amine *via* nucleophilic substitution and reductive amination. However, even after extensive screening, no products could be isolated at all (Scheme 26).

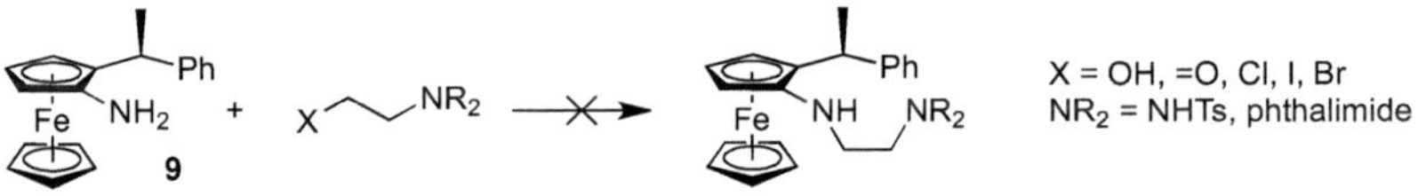

**Scheme 26:** The target ligand could not be prepared *via* alkylation or reductive amination.[1]

Thus, the reactivity of amine **9** in other reactions was investigated. Much to our surprise, Hartwig-Buchwald cross-coupling[85, 86] with 2-bromopyridine yielded the corresponding aminopyridine **22** in 95% yield. Also, the amine could be reacted with *N,N*-dimethylformamide dimethylacetal to give the corresponding amidine **23** in quantitative yield (Scheme 27).

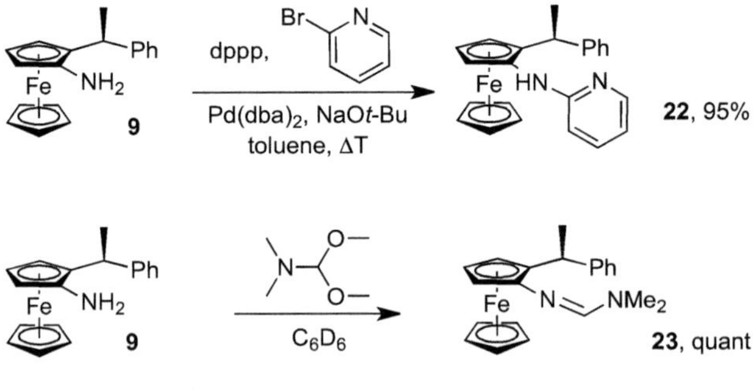

**Scheme 27:** Successful functionalization of **9**: Hartwig-Buchwald amination (top) and amidine formation (bottom).

Especially the formation of the amidine was not expected, as the amine had not shown any evidence to form imines when it was tried to react it with aliphatic aldehydes. As a consequence, reductive amination of aryl aldehydes was attempted. Condensation of ferrocenyl amine **9** and picolinaldehyde and subsequent reduction with sodium borohydride gave the corresponding amine **25** in 94% yield. When adapting this procedure to the reaction with 2-(diphenylphosphino)benzaldehyde, however, the attempted reduction of the intermediate imine **26** led to a mixture of products; reduction to amine **27** could not be confirmed (Scheme 28).

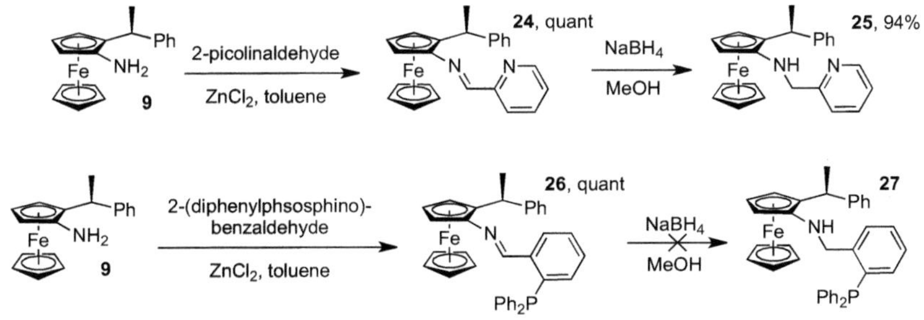

**Scheme 28:** Condensation of amine **9** with aryl aldehydes was successful, subsequent reduction gave ambiguous results.

Inspired by the fact that ferrocenylimines could be formed and reduced to the corresponding amines, it was tried to prepare an *en*-type ligand. Reacting glyoxal with two equivalents of amine **9** gave an inseparable mixture of products, tentatively the corresponding *s-cis* and *s-trans* diimines **28** and a small amount of starting material. Nevertheless, it was tried to reduce the mixture, as full reduction should yield the corresponding diaminoethane. Again, a mixture of products was obtained which could not be separated (Scheme 29).

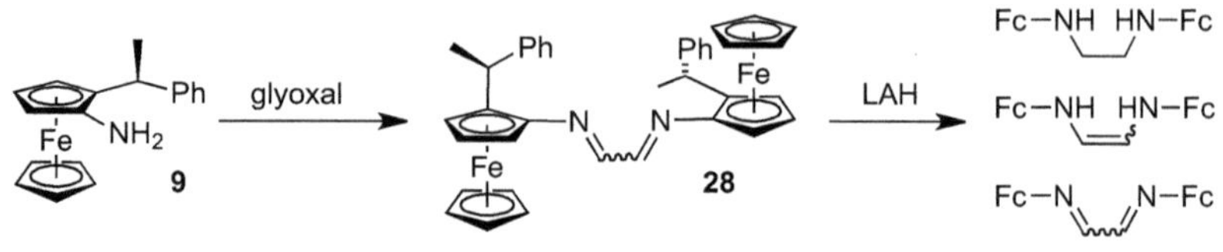

**Scheme 29:** Attempted synthesis of an *en*-type ligand gave inseparable product mixtures.

In addition it was tried to condensate acetylacetone with two equivalents of ferrocenyl amine **9**. Detailed results and ensuing synthetic work is discussed in Section 2.4.

### 2.3.3 Complex Synthesis

Even though the projected tridentate ligand could not be prepared, complexes containing the primary amine **9**, aminopyridine **22** or picolinamine **25** could also act as bifunctional catalyst in transfer hydrogenation.

It was tried to coordinate all three compounds as neutral and anionic ligands, mainly to ruthenium(II) precursors. Because the functionalized amines **22** and **25** can also act as bidentate ligands without a metal-arene interaction, coordination to both iridium(I) and iridium(III) was tried as well. Representative attempts are summarized in Table 2.3.

Coordination of the primary amine **9** as a neutral ligand to the commonly used dinuclear precursor ruthenium(II) *para*-cymene dichloride was tried first (entry 1). As no complex formation could be confirmed, precursors with less electron-rich arenes were used (entries 2–6). Eventually, the amine was deprotonated with *n*-BuLi before addition of the precursor (entries 9 and 10). Also, precursors with more labile ligands, such as DMSO or cyclooctadiene were tried (entries 7 and 8). No coordination compounds were formed in any case.

Similarly, the secondary amines were first tried as neutral (entries

11 and 12) and then as anionic ligands to ruthenium(II) precursors (entries 13 and 14). Coordination to iridium precursors was only tried with anionic ligands (entries 15–18). Again, no complex formation was observed.

Table 2.3: Unsuccessful coordination attempts.

| entry | [M] | ligand | base | solvent | temp/[°C] | time/[d] |
|---|---|---|---|---|---|---|
| 1 | [Ru($p$-cymene)Cl$_2$]$_2$ | **9** | - | toluene-d$_8$ | 100 | 3 |
| 2 | [Ru(Ph)Cl$_2$]$_2$ | **9** | - | toluene-d$_8$ | 100 | 3 |
| 3 | [Ru(mmb)Cl$_2$]$_2$ | **9** | - | DMSO-d$_6$ | 70 | 5 |
| 4 | [Ru(mmb)Cl$_2$]$_2$ | **9** | - | CD$_3$CN | 70 | 3 |
| 5 | [Ru(etb)Cl$_2$]$_2$ | **9** | - | CD$_2$Cl$_2$ | 50 | 5 |
| 6 | [Ru(etb)Cl$_2$]$_2$ | **9** | - | CD$_3$OD | 50 | 5 |
| 7 | [Ru(dmso)$_4$Cl$_2$]$_2$ | **9** | - | toluene-d$_8$ | 100 | 3 |
| 8 | [Ru(cod)Cl$_2$]$_2$ | **9** | - | toluene-d$_8$ | 100 | 3 |
| 9 | [Ru(Ph)Cl$_2$]$_2$ | **9** | $n$-BuLi | THF-d$_8$ | 0–60 | 3 |
| 10 | [Ru(etb)Cl$_2$]$_2$ | **9** | $n$-BuLi | THF-d$_8$ | 0–60 | 3 |
| 11 | [Ru(etb)Cl$_2$]$_2$ | **22** | - | toluene-d$_8$ | 60 | 3 |
| 12 | [Ru(etb)Cl$_2$]$_2$ | **25** | - | toluene-d$_8$ | 60 | 3 |
| 13 | [Ru(etb)Cl$_2$]$_2$ | **22** | $n$-BuLi | THF-d$_8$ | 0–60 | 3 |
| 14 | [Ru(etb)Cl$_2$]$_2$ | **25** | $n$-BuLi | THF-d$_8$ | 0–60 | 3 |
| 15 | [Ir(coe)$_2$Cl]$_2$ | **22** | $n$-BuLi | THF-d$_8$ | 0–60 | 3 |
| 16 | [Ir(Cp*)Cl$_2$]$_2$ | **22** | $n$-BuLi | THF-d$_8$ | 0–60 | 3 |
| 17 | [Ir(coe)$_2$Cl]$_2$ | **25** | $n$-BuLi | THF-d$_8$ | 0–60 | 3 |
| 18 | [Ir(Cp*)Cl$_2$]$_2$ | **25** | $n$-BuLi | THF-d$_8$ | 0–60 | 3 |

## 2.4 β-Diketiminato Ligands

β-Diketimines synthesized *via* condensation of acetylacetone with two equivalents of a primary amine have been known since the early 1920s; the structural moiety is commonly referred to as *HNacNac*. In a publication primarily focussed on the color of dyes, Scheibe described the reaction product of aniline and acetylacetone as being prone to hydrolysis. However, the corresponding hydrochloride as well as a dichloro zinc complex were reported (Scheme 30).[87]

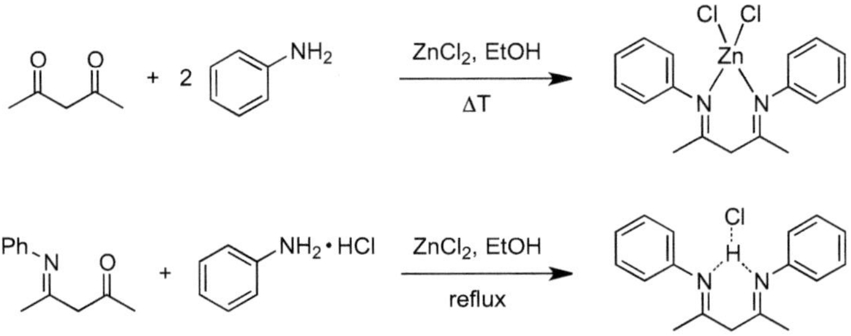

**Scheme 30:** First synthesis of NacNac ligands and complexes as reported by Scheibe.[87]

During the following decades, the synthesis of NacNac type ligands was not explored very thoroughly. Most syntheses gave either the corresponding salts in two steps,[88, 89] or the corresponding metal complexes in one to four steps.[89, 90] A metal-free synthesis was only published four decades later by Dorman (Scheme 31). It was also pointed out that the NacNac system may be considered as a non-classical aromatic system due to its planarity and the presence of hydrogen bonding.[91]

New interest in this class of ligands arose mainly due to widespread successful applications of ligands bearing two imine moieties in catalysis. Noteworthy examples are Pfaltz' semicorrins,[92]

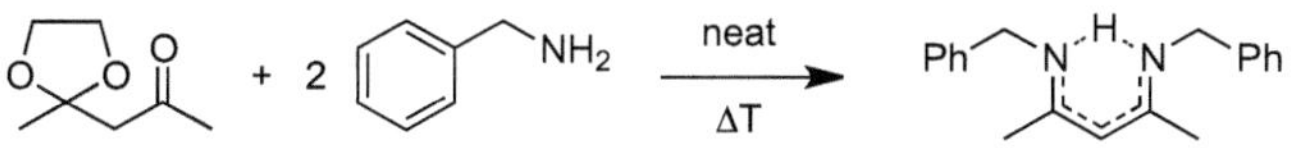

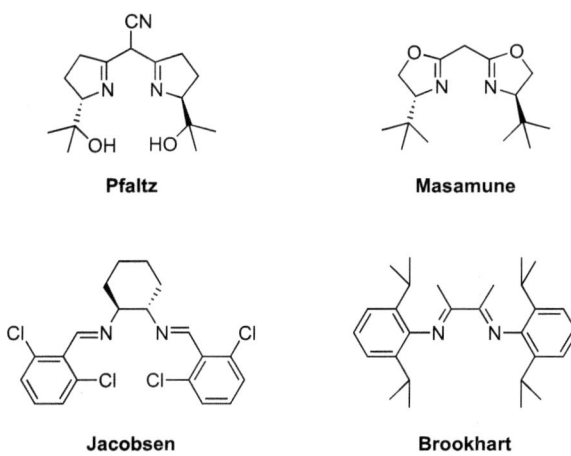

**Scheme 31:** Synthesis of N,N'-dibenzyl NacNac as reported by Dorman.[91]

the bis(oxazoline) family developed by Masamune,[93] Pfaltz,[53] Evans[94] and Corey,[95] the 1,2-diaminocyclehexane-based ligands presented by Jacobsen,[96] or the polymerization catalysts of Brookhart and co-workers [97] (Figure 10).

**Figure 10:** Successful diimine ligands.[92, 93, 96, 97]

## 2.4.1 Chiral NacNac Ligands

In contrast to the wide variety of chiral β-diimine ligands available, only one chiral NacNac ligand has been reported so far: bis-N,N'-(2-phenylethyl)-2,4-diiminopentane. A two-step synthesis was first published by Buch *et al.* [98, 99] and a one-pot route was reported shortly thereafter by Schaper and co-workers [100] (Scheme 32).

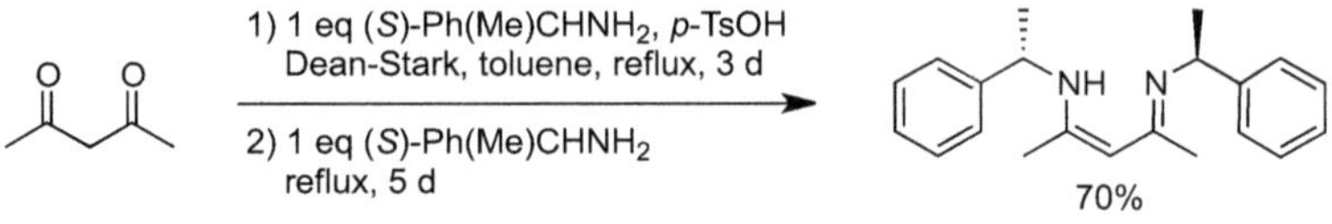

**Scheme 32:** One-pot synthesis of a chiral NacNac ligand as reported by Schaper and co-workers.[100]

## 2.4.2 Applications of NacNac Complexes

β-Diiminates have so far been reported for forty-nine out of the sixty-three natural metals. Often, the ligands are used to stabilize metal ions in unusual oxidation states or coordinatively unsaturated metal centers. Catalytic applications of NacNac complexes include polymerization, hydrogenation, hydrosilylation, aziridination, C-H activation, metathesis, and inter- and intramolecular hydroamination reactions. As coherent and recent reviews are available,[101, 102] only a handful of highlights are presented here.

### Intramolecular Hydroamination

Various NacNac complexes found applications in intramolecular hydroamination reactions. Among these the ones containing earth alkali or an early transition metal or as well a lanthanide are the most commonly used (Scheme 33).

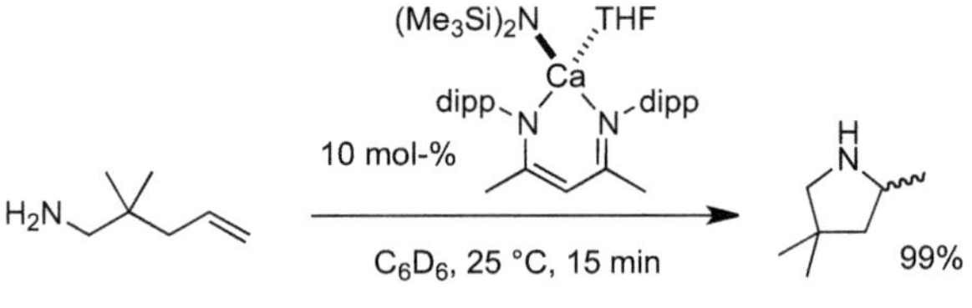

**Scheme 33:** Calcium-mediated intramolecular hydroamination as reported by Hill and co-workers.[103]

## Dinitrogen Complexes

Dinitrogen complexes have proven to be highly reactive catalyst precursors. The iridium(I) and rhodium(I) NacNac complexes were used as hydrogenation catalysts, [104, 105] the dinuclear chromium(I) complex may be used for the activation of small molecules[106] (Scheme 34).

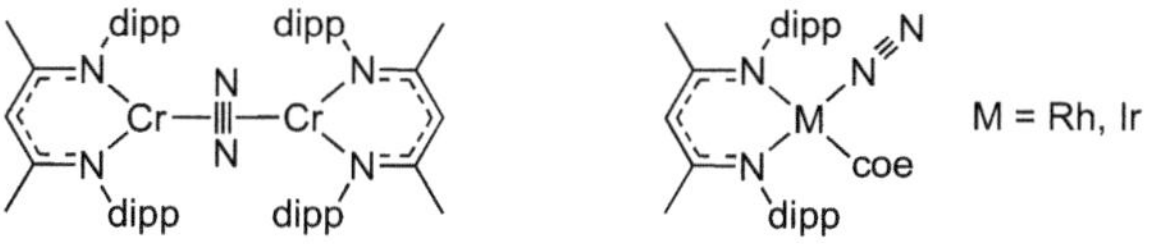

**Scheme 34:** Dinitrogen complexes of $Cr^I$,[106] $Rh^I$,[105] and $Ir^I$.[104]

## Gold(I)-Catalyzed Aerobic Oxidation of Alcohols

As Section 3.3 of this thesis deals with gold(I) complexes it is only natural to mention that NacNac ligands have found application in gold(I) catalysis. In 2005 Shi and co-workers reported a versatile method for the oxidation of alcohols using *in situ* prepared gold(I) NacNac catalysts[107] (Scheme 35).

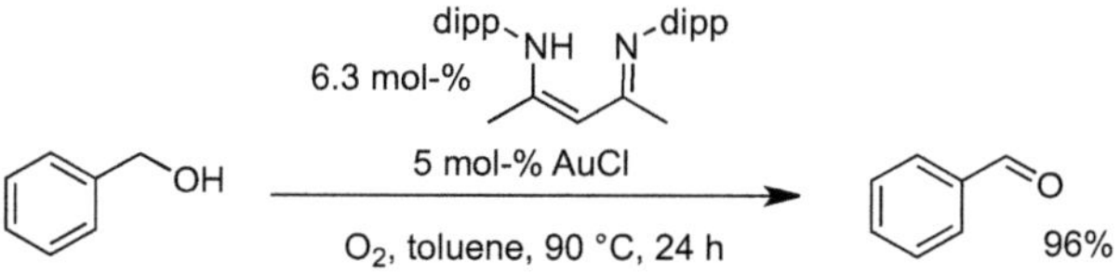

**Scheme 35:** Aerobic oxidation of alcohols as reported by Shi and co-workers.[107]

## $\eta^6$-Arene $\beta$-Diketiminato-Ruthenium Complexes

In 2007, Dyson and co-workers reported novel ruthenium(II)-NacNac complexes[108] which catalyze alkene hydrogenation[109] and were found to be more cytotoxic towards ovarian cancer cells than *cis*-platin[110] (Scheme 36).

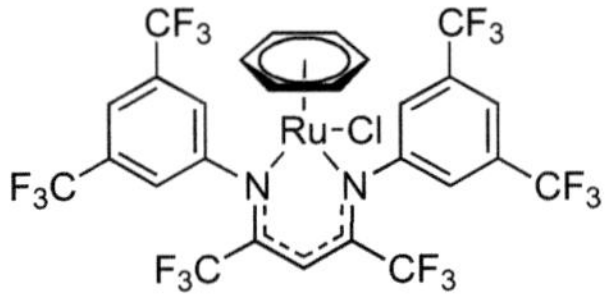

**Scheme 36:** A highly cytotoxic ruthenium(II)-NacNac complex as reported by Dyson and co-workers.[110]

**Stereoselective Applications**

As there is only one chiral NacNac ligand known in the literature,[98] not many applications have been published. The only report on a catalytic application of a zinc(II) complex containing this ligand discusses its use in lactide polymerization. However, no enantioselectivity was observed[111] (Scheme 37).

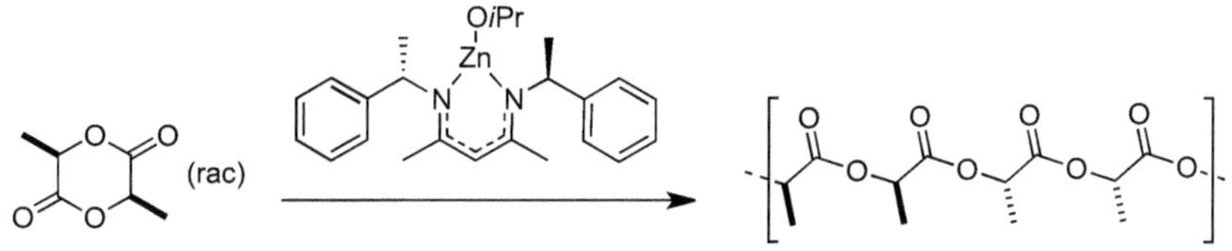

**Scheme 37:** Polymerization of *rac*-lactide as reported by Schaper and co-workers.[111]

## 2.4.3 Diferrocenyl-NacNac

With aminoferrocene **9** in hand, the synthesis of the corresponding NacNac ligand was not expected to take more than two steps. In a first attempt, a solution of acetyl acetone, two equivalents of amine **9** and one equivalent of hydrochloric acid in ethanol were heated at reflux for two days, giving the target product **29** in low yields. Screening led to the use of higher-boiling toluene as a solvent, which allowed the removal of nascent water by using a Dean-Stark trap. Due to its volatility, hydrochloric acid was replaced by *para*-toluensulfonic acid. In addition it was observed that best results were obtained when the amine was added in two batches (Table 2.4).

Table 2.4: Synthesis of diferrocenyl-NacNac **29**.

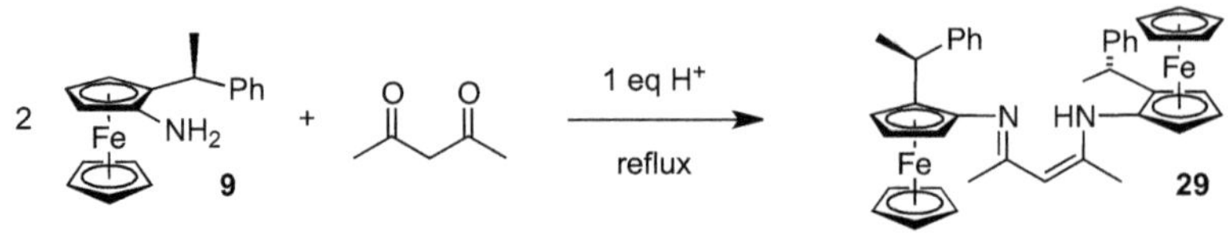

| entry | solvent | acid | time /[h] | scale /[mg][a] | yield /[%] |
|---|---|---|---|---|---|
| 1 | EtOH | HCl | 44 | 180 | 24 |
| 2 | EtOH | HCl | 44 | 500 | 42 |
| 3 | i-PrOH | HCl | 20 | 500 | 32 |
| 4 | toluene[b] | p-TsOH · H$_2$O | 20 | 360 | 45 |
| 5 | toluene[c] | p-TsOH · H$_2$O | 5+40 | 2000 | 75 |
| 6 | toluene[c] | p-TsOH · H$_2$O | 5+40 | 5000 | 81 |

[a]Values correspond to the amount of **9**. [b]MS 4 Å. [c]Dean-Stark trap.

In 2011, more than one year after our last synthesis of HNac-Nac **29**, Stephan and co-workers published a two-step synthesis of bis(ferrocenyl)NacNac with an overall yield of 50% (Scheme 38).[112] In a first step, only one equivalent of aminoferrocene was reacted with acetylacetone in dichloromethane at room temperature. The ferrocenylimine-$\beta$-ketone intermediate was isolated and the HNacNac was prepared in a second step by addition of another equivalent of amine and heating to reflux for four days. Interestingly, it was claimed that a one-pot synthesis or the presence of acid gave only inseparable mixtures of products.

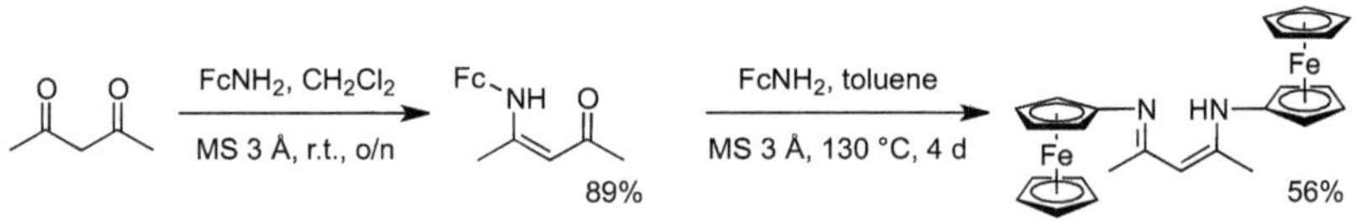

**Scheme 38:** Synthesis of bis(ferrocenyl)NacNac as reported by Stephan and co-workers.[91]

### 2.4.3.1 Structural Features

First hints on the stucture could be extracted from the $^1$H NMR data. Two singlet signals at 11.22 and 4.37 ppm and the absence of a CH$_2$-group indicated that the imine-enamine tautomer is favored in solution. The same was observed in the crystal structure (Figure 12). The NacNac unit is almost completely delocalized, with its corresponding bond lenghts nearly converging. Examination of the bond lenghts showed slightly shorter bonds alternating with slightly longer ones, which would locate the amino group at N2. Before the final refinement, it was tried to deduce the location of the hydrogen atom from an electron density plot. As shown in Figure 11, protonation of N2 is in better agreement with the experimental results.

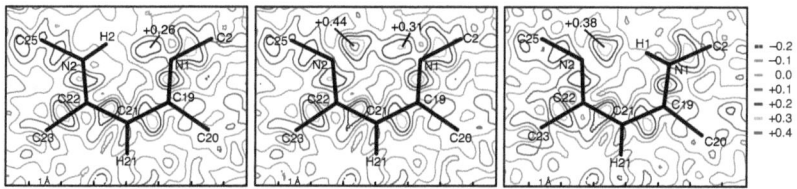

**Figure 11:** Bruker XP 5.1 $F_o$–$F_c$ electron density contour plot. Mean square plane calculated through N1–C19–C21–C22–N2. Values are given in e Å$^{-3}$. Left: min −0.3, mean +0.03, max +0.49. Middle: min −0.32, mean +0.04, max +0.47. Right: min −0.26, mean +0.03, max +0.40.

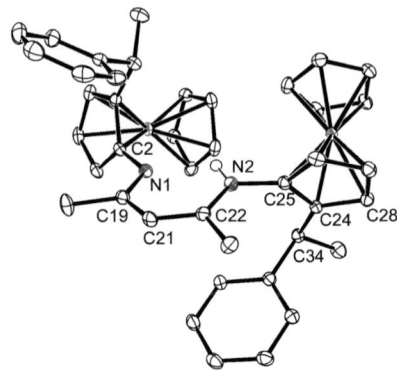

**Figure 12:** ORTEPIII representation of HNacNac **29**. Except for the amino proton, hydrogen atoms are omitted for clarity, thermal ellipsoids are set to 50% probability. Selected bond lengths [Å], bond and torsion angles [°]: C2–N1 1.415(2), N1–C9 1.328(2), C19–C21 1.411.(2), C21–C22 1.398(2), C22–N2 1.347(2), N2–C25 1.416(2), C2–N1–C19 121.21(14), N1–C19–C21 120.68(15), C19–C21–C22 126.57(15), C21–C22–N2 121.48(15), C22-N2–C25 120.44(14), C1–C11–C12 111.89(14), C1–C11–C13 110.69(13), C12–C11–C13 111.34(14), C24–C34–C35 111.92(14), C24–C34–C36 110.23(13), C35–C34–C36 111.28(14), N1–C19–C21–C22 2.0(3), N2–C22–C21–C19 3.5(3), N1–C19–C22–N2 4.7(3), C1–C2–N1–C19 112.87(18), C24–C25–N2–C22 103.15(19), C5–C1–C11-C12 −23.6(2), C5–C1–C11–C13 101.21(19), C1–C11–C13–C14 108.37(18), C28–C24–C34–C35 −10.1(2), C28–C24–C34–C36 114.35(18), C24–C34–C35–C37 114.90(17).

## 2.4.3.2 Complex Syntheses

Although β-diketimines can, similarly to e. g. bisoxazolines, act as neutral ligands, only few examples of complexes with neutral HNacNac ligands exist. They usually are used as anionic ligands. Intuitively, one may think of three main synthetic strategies:

- Deprotonation of the ligand with bases, such as $n$-BuLi or KH to form the corresponding alkali salts which can then be reacted with a metal halide.

- Use of metal precursors with Brønsted basic ligands.

- Oxidative addition of the N-H bond to give the corresponding hydride.

While the first strategy needs an additional reagent to deprotonate the ligand, e. g. $n$-BuLi, KH or NaH, it can then react with the most common metal precursors and is very convenient. The other two methods need more reactive metal precursors which may also react with the ligand in an unwanted fashion. Also, they are intrinsically more difficult to handle.

It was first attempted to deprotonate the ligand before coordination. In order to control the stoichiometry, a $n$-BuLi solution of low concentration was freshly prepared from 1-chlorobutane and lithium in hexane. Its concentration was determined to be 0.11 M $via$ titration with diphenyl ditelluride.[113] Although a reaction with the base occurred—visualized by a color change from orange to bright red—no coordination to the metal precursors could be confirmed (Table 2.5).

Table 2.5: Attempted preparation of NacNac-complexes after deprotonation.

| entry | base | metal precursor | solvent | comment |
|---|---|---|---|---|
| 1 | $n$-BuLi | Zr(NMe$_2$)$_4$ | THF | no reaction |
| 2 | $n$-BuLi | [Cu(MeCN)$_4$]ClO$_4$ | THF | dark precipitate |
| 3 | KO$t$-Bu | CuCl$_2$ | THF | no reaction |
| 4 | $n$-BuLi | [Ir(cod)Cl]$_2$ | THF | no reaction |
| 5 | $n$-BuLi | [Pd(allyl)Cl]$_2$ | THF | dark precipitate |
| 6 | $n$-BuLi | (Me$_2$S)AuCl | THF | Au mirror |

Due to the relatively high acidity of the HNacNac moiety, even metal acetates are basic enough to deprotonate the ligand without addition of another base.[114, 115] Commonly used precursors are alcoholates and amido complexes, such as titanium(IV) $iso$-propoxide or tetrakis(dimethylamido)zirconium(IV). A drawback in all of the mentioned approaches is that the protonated ligands from the precursor remain in the reaction mixture. Metal alkyls or -hydrides are more convenient because the formed gases may be easily removed from the reaction mixture. All attempts to form complexes according to this strategy are presented in Table 2.6.

Table 2.6: Attempted preparation of NacNac-complexes *via* deprotonation.

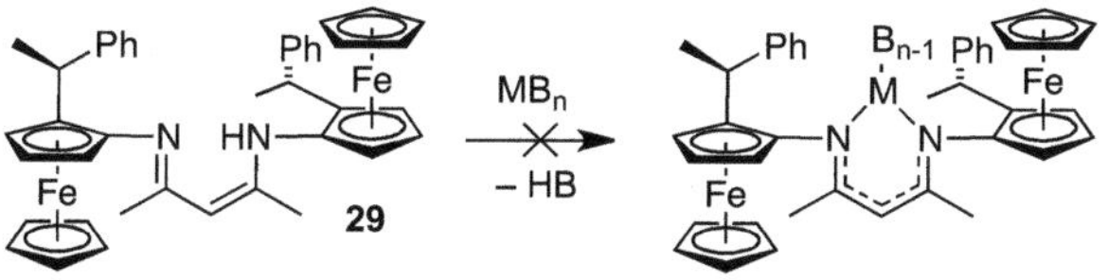

| entry | metal precursor | solvent | comment |
|---|---|---|---|
| 1 | Ti(O*i*-Pr)$_4$ | THF | decomp. |
| 2 | Zr(NMe$_2$)$_4$ | THF | see Fig. 13 |
| 3 | Pd(OAc)$_2$ | acetone | decomp. |
| 4 | ZnMe$_2$ | toluene | decomp. |
| 5 | ZnEt$_2$ | THF | decomp. |

As it was mentioned before, Stephan and co-workers reported the synthesis of bis(ferrocenyl)NacNac ligands after we had discontinued our project.[112] In their publication, they also reported the reaction of their ligands with dimethylzinc, which yielded the corresponding zinc(II) complexes. When attempting to abstract the methyl group with a trityl cation in order to generate a coordinatively unsaturated metal center, they observed alkylation of the ligand, as shown in Scheme 39. They suggest that this reactivity is caused by the steric demand of the ferrocenyl substituents.

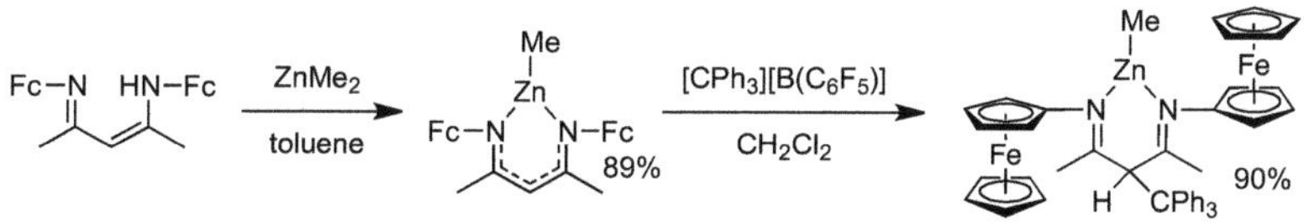

**Scheme 39:** Preparation and reactivity of [ZnMe(Fc$_2$NacNac)] as reported by Stephan and co-workers.[112]

In contrast to the results from Stephan, none of our attempts yielded products characterizable by $^1$H NMR. Usually, a dark solution containing a black precipitate was obtained. Only in the case of $Zr(NMe_2)_4$ (entry 2, Table 2.6), a small number of orange crystals could be isolated. Again, $^1$H NMR data were inconclusive, and the quality of the crystals ($R_{int} = 0.18$) did not allow to refine the structure completely. In addition to the low quality of the dataset, the asymmetric unit contains four NacNac units, a number of THF molecules, and possibly even $Zr(NMe_2)_n$ units, which could not be confirmed unambiguously. The unit cell content is shown in Figure 13. Due to the uncertain nature of the smaller fragments, only the NacNac molecules are shown.

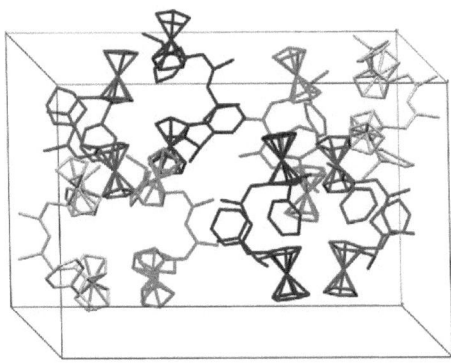

**Figure 13:** Change of conformation after reaction with $Zr(NMe_2)_4$. MERCURY-generated representation of the unit cell content. Symmetry equivalent molecules are of the same color. Hydrogen atoms were not added. Solvent molecules and unassignable moieties are ommitted for clarity.

Even though this picture only shows a low-quality stucture of a minor product of a reaction which was not reproduced, it gives an important clue why coordination attempts were not successful. Instead of coordinating in a bidentate fashion, the conformation of the Nac-

Nac moiety was changed drastically, possibly due to the steric demand of the ferrocenyl substituents, which make the coordination site much less accessible compared to common 'bulky' dipp$_2$NacNac ligands or the zinc(II) complex reported by Stephan and co-workers (Figure 14). As a result, not only the formation of the envisioned complexes, but also efficient catalytic applications of such compounds seem unlikely.

When examining the unassignable peaks of the residual electron density maxima, several can be found close to the nitrogen atoms. Their location could indicate an $\eta^3$-azaallyl-type coordination, as speculatively shown in Scheme 40. It has to be pointed out, however, that the NacNac moieties of all four molecules of the asymmetric unit show different geometries and bond lenghts, therefore no cogent conclusions about possible coordination modes may be drawn.

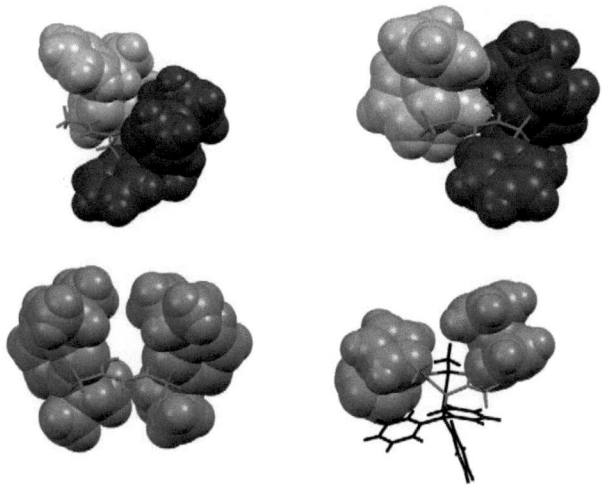

**Figure 14:** MERCURY visualisation of the steric demand of the ferrocenyl units of **29** (top), the popular 'bulky' dipp$_2$NacNac ligand[116] (bottom left) and a derivative of Stephan's bis(ferrocenyl)NacNac ZnMe complex[112] (bottom right). Spacefill surfaces correspond to van der Waals radii.[117, 118]

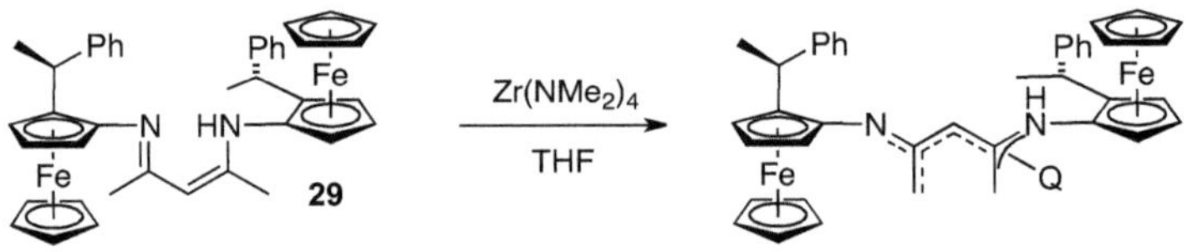

**Scheme 40:** The reaction of NacNac **29** with Zr(NMe$_2$)$_4$ possibly leads to aza-allylic coordination.

To complete the strategies discussed above for the synthesis of metal complexes containing an anionic NacNac ligand, it was tried to coordinate ligand **29** to nickel(II) *via* oxidative addition. When bis(cyclooctadiene)nickel(0) and the ligand were dissolved in deuterated benzene, only the formation of a black ferromagnetic precipitate—presumably elemental nickel—was observed (Scheme 41).

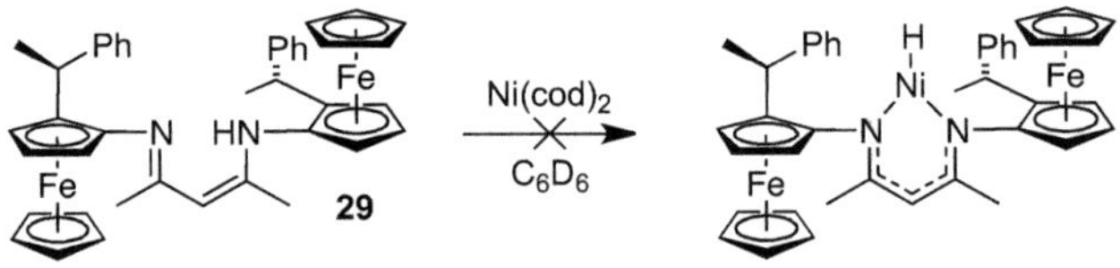

**Scheme 41:** Attemptetd synthesis of NiH(**29**) *via* oxidative addition.

It was also tried to use HNacNac **29** as a neutral ligand. Due to its solubility and relatively small size, it was decided to use dimethylsulfidecopper(I) bromide as metal precursor. No reaction was observed (Scheme 42).

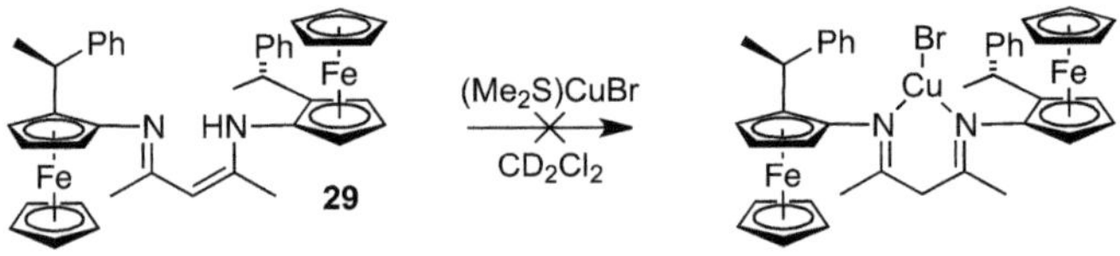

**Scheme 42:** Attempted coordination of neutral **29** to copper(I) bromide.

### 2.4.3.3 Iron(II)-Catalyzed Hydrosilylation of Acetophenone

During the Ph. D.-thesis of Michelle Flückiger, NacNac-ligand **29** was probed for its potential in iron(II)-catalyzed hydosilylation of ketones.[119] While the ligand proved to afford active catalysts, as shown by a three times higher conversion in comparison to the ligand-free reaction, stereoselectivity was low. Taking into account that the active species was only formed *in situ* and a bidentate coordination of the ligand had not been observed to any metal (*vide supra*), it seems reasonable to assume that NacNac **29** acts as a monodentate ligand with little steric influence. Condensed results are given in Table 2.7.

Table 2.7: Hydrosilylation of acetophenone.[119]

| entry | ligand | conversion /[%] | R:S |
|---|---|---|---|
| 1 | – | 36.6 | *rac* |
| 2 | **29** | 90.1 | 47.5 : 52.5 |
| 3 | **30** | quant | 72.5 : 27.5 |

## 2.5 Ferrocenyl Ureas

### 2.5.1 Introduction

While Wöhler's 1828 article *Ueber künstliche Bildung des Harnstoffs*[120] marked the beginning of a whole new era in organic synthesis and led to a golden age of natural product synthesis, urea and its derivatives were mostly neglected due to their low reactivity. With the emergence of organocatalysis, and non-covalent organocatalysis in particular, the ability of urea and its derivatives to form multiple hydrogen bonds sparked new interest in this class of compounds. Consequently, urea functionalities are now used in several fields, including supramolecular chemistry, anion recognition and polymer chemistry.[121, 122]

### 2.5.2 Urea Derivatives in Organocatalysis

Since the first report on the catalytic application of a symmetrical *N,N'*-diaryl urea by Curran *et al.* in 1994,[123] countless derivatives have been designed to fit varying steric and electronic demands. The bonding affinity of the urea moiety to the substrates may be tuned by altering the substituents on the nitrogen atoms. Alternatively, thiourea were also used, which made an even wider ranges of catalysts easily accessible.

Concerning asymmetric organocatalysis, the most beneficial feature of urea are the two parallel N–H bonds which are well suited to doubly bind to hydrogen-bond acceptors (Figure 15). This leads to well-defined rigid structures, in which the influence of bulky chiral substituents can be maximized.

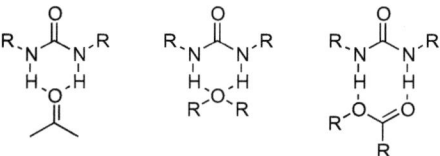

**Figure 15:** Double hydrogen bonding of urea to ketones, ethers and esters.

### 2.5.3 Ferrocenyl Ureas

In the course of optimizing the synthesis of ferrocenylamine **9**, it was tried to shortcut the Cbz-protected amine by quenching the intermediately formed isocyanate with water instead of benzylic alcohol (Scheme 43). After all, the free amine was air stable and the need for a protecting group was not as immediate as in the case of Bertogg's sensitive substances;[77] a one-pot procedure would avoid one day of work and possibly enhance the overall yield significantly.

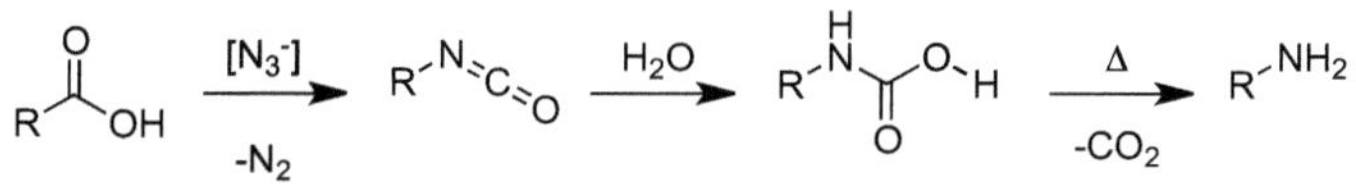

**Scheme 43:** One-pot conversion of carboxylic acids to primary amines *via* Curtius rearrangement.

Following the same procedure as for the preparation of carbamate **8**, but quenching with water, gave an orange cottonish solid in quantitative yield. At first glance, $^1$H NMR seemed to confirm a success, however, the physical appearance of the product left doubts. While other analytic methods indicated that not the amine but something else was formed after all, mass spectroscopy gave the crucial hint on the actual composition of the isolated product: the mass of the main fragment was attributed to twice the mass of the amine plus a carbonyl group.

$$636 = 2 \times (M_m(\mathbf{9})) - H_2 + 28 \, (\hat{=} CO)$$

Thus, bis(ferrocenyl)urea had been formed. Apparently, the initial idea of forming the free amine *in situ* worked, but due to its increased nucleophilicity, the product reacted with the isocyanate to yield urea **31** (Scheme 44). In the wake of this result, a selection of monoferrocenyl ureas were prepared by quenching with primary aromatic amines (Table 2.8).

The known tendency of urea derivatives to form strong hydrogen bonds became apparent during the preparation of analytically pure samples. When the products were purified by flash chromatography using mixtures of *n*-hexane and ethyl acetate as eluents, the isolated solid products contained about one equivalent of ethyl acetate. This residual solvent had to be removed tediously by repeated azeotropic distillation; prolonged exposure to high or ultra high vacuum ($10^{-9}$ mbar) was insufficient.

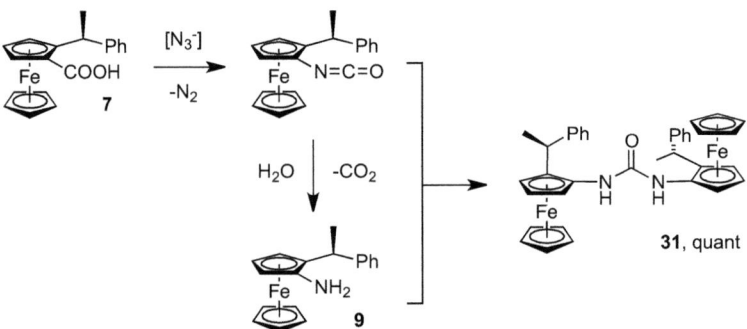

**Scheme 44:** Instead of the free amine **9**, bis(ferrocenyl) urea **31** was isolated when quenching the reaction with water.

Table 2.8: Preparation of ferrocenyl ureas.

| entry | nucleophile | R | product | yield[a] /[%] |
|---|---|---|---|---|
| 1 | $H_2O$ | 2-(1-phenylethyl)ferrocen-1-yl | **31** | quant |
| 2 | aniline | Ph | **32** | 69 |
| 3 | o-anisidine | 2-methoxyphenyl | **33** | 73 |
| 4 | 6-methylpyridin-2-amine | 6-methylpyridin-2-yl | **34** | 84 |

[a]Yields are unoptimized, reactions were carried out once.

### 2.5.3.1 Application of Ferrocenyl Ureas in Organocatalysis: Preliminary Results

When these novel ureas were prepared, the focus of this thesis had already shifted to phosphinoferrocenes, which will be discussed in Chapter 3. Therefore, the application of ferrocenyl ureas as chiral organocatalysts was not examined in depth and the presented catalyses only represent preliminary results.

#### 2.5.3.1.1 Michael Addition of Diethyl Malonate to Nitrostyrene

A variety of urea and thiourea derivatives have been shown to catalyze Michael[124–127] and nitro-Michael[128–136] additions, among them also the valine-based (thio)urea published by Pedrosa and co-workers.[137] Compared to most other chiral urea-based catalysts, the latter are structurally quite simple compounds. Nonetheless, they reached high enantioselectivities (Table 2.9, entries 1 and 2). As our ferrocenyl ureas are no fancy compounds either, nitro-Michael addition appeared to be a good choice for first trials.

Table 2.9: Enantioselective Michael addition.

| entry | cat | temp /[°C]    | time /[h] | yield /[%] | er   |
|-------|-----|---------------|-----------|------------|------|
| 1     | 35  | r. t.         | 22        | 93         | 93:7 |
| 2     | 36  | −18           | 44        | 90         | 97:3 |
| 3     | 33  | r. t. to reflux | 120     | 0          | -    |

When using the same conditions Pedrosa and co-workers used for their urea catalyst **35**, no product was detected after 24 h. Therefore, the temperature was raised stepwise by 20 °C per day. Even after one day at reflux—which gives a total reaction time of six days—no product was formed.

### 2.5.3.1.2 Indium-Mediated Allylation of Acylhydrazones

It was decided to examine the influence of the novel ferrocenyl urea on indium-mediated allylation of acylhydrazones. A stereoselective version of this reaction has been published by Jacobsen *et al.* (Table 2.10, entries 1 and 2).[138] Because of the frustrating experiences with the steric bulk of bis(ferrocenyl)NacNac **29**, only mono(ferrocenyl) ureas were tested. As Jacobsen's best result was obtained using a urea with an oxygen-containing directing group, *ortho*-methoxyphenyl urea **33** was used. In order to assess whether the methoxy group influences the reaction at all, phenyl urea **32** was tried as well.

Table 2.10: Asymmetric allylation of acylhydrazones.

| entry | catalyst | yield /[%] | er[a] |
|---|---|---|---|
| 1[b] | 37 | 79 | 69:31 |
| 2[b] | 38 | 87 | 95:5 |
| 3 | 33 | 44 | 68:32 |
| 4 | 32 | 46 | 66:34 |

[a] Absolute configuration was not determined. [b] −20 °C, 19 h; Yields determined by HPLC.[138]

Not too surprisingly, it seems that the ferrocenyl ureas could not keep up with the performance of Jacobsen's much more sophisticated catalysts. It has to be noted, however, that in order to get reasonable conversions, the catalyses were carried out at −10 °C. This change in temperature compared to Jacobsen's protocol may well account for a drop in enantioselectivity. Also, our isolated yields differ drastically from NMR-yields, which were nearly quantitative. This is not due to the sensitivity of the products but to an impurity which required triple flash chromatography. Taking these facts into account, both ferrocenyl urea derivatives perform at least as well as the established thiourea **37**. Also, the differences between the two catalysts are very small and lie within the experimental error.

## 2.6 Additional Investigations

### 2.6.1 A Longer Tether

At a later stage of the project, elongation of the tether by one carbon unit was taken into consideration. While elongation of the side chain would need various adjustments in the synthesis, inserting an additional carbon between the Cp-ring and the amine would alter the route only after The introduction of the aryl group. An obvious approach is the synthesis *via* an aldehyde, which may then be coupled with a primary amine by reductive amination. As shown in Scheme 45, ferrocenylaldehyde **39** was prepared quantitatively by quenching lithiated bromide **14** with an excess of *N,N*-dimethylformamide.

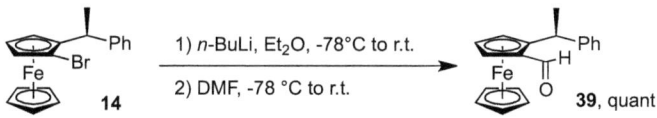

**Scheme 45:** Synthesis of ferrocenylaldehyde **39**.

Aldehyde **39** was isolated as bright red microcrystals. X-ray quality crystals were obtained by slow evaporation of a saturated solution of the compound in dichloromethane. An ORTEPIII-representation of the refined structure is shown in Figure 16. The conformation of the sidechain is very similar to the one of the parent bromide; differences in bond angles are not significant.[†] The carbonyl group is slightly tilted towards the iron; the angle between the cyclopentadienyl plane and the newly formed bond was calculated to 5.3(3)°. The carbonyl group itself is rotated away from the iron center by 2.7(3)°, leaving the oxygen in plane with the cyclopentadiene.

As the focus of this thesis shifted towards phosphinoferrocenes, this approach towards a coordinating ligand was not examined any

---

[†]i. e. $|A - B| < 3\sqrt{\sigma^2(A) + \sigma^2(B)}$.

further.

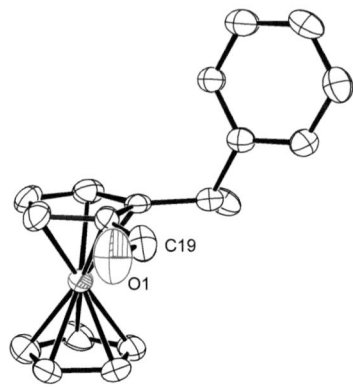

**Figure 16:** ORTEPIII-representation of ferrocenylaldehyde **39**. Hydrogen atoms are omitted for clarity, thermal ellipsoids are set to 50% probability. Selected bond lengths [Å], bond and torsion angles [°]: O1–C19 1.203(3), C2–C19 1.457(4), O1–C19–C2 123.8(3), C1–C2–C19 125.3(3), C1–C11–C12 113.3(2), C1–C11–C13 111.9(2), C12–C11–C13 109.9(2), O1–C19–C2–C1 177.3(3), C1–C11–C13–C15 137.5(3), C5–C1–C11–C12 −19.4(4), C5–C1–C11–C13 105.5(3).

## 2.7 Conclusions

At the beginning of this thesis, the synthesis of the primary ferrocenyl amine **9** was substantially improved to an overall yield of 67% over five steps and the potential modularity of the ligand was shown by introducing a total of seven different aryl groups on the sidechain.

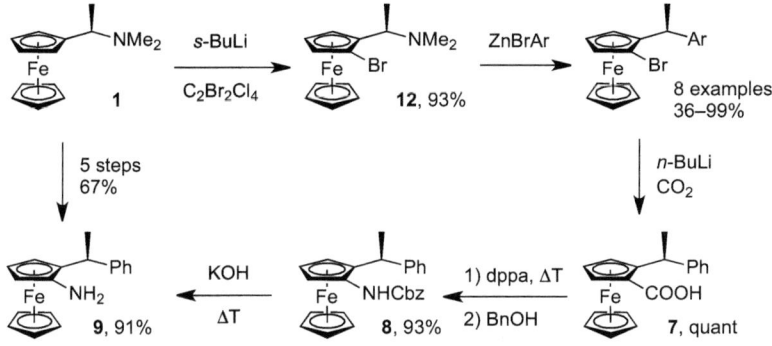

Pursuing the inital goals of the thesis, the author found his 'Waterloo' in the final step of the ligand synthesis due to the erratic reactivity of the primary ferrocenylamine **9**. Most commonly attempted pathways turned out to be dead ends due to no product formation or inseparable mixtures.

Because ferrocenylamine **9** could not be coordinated to ruthenium(II) precursors, the focus shifted to the synthesis of bidentate ligands. Bis(ferrocenyl)NacNac **29** was synthesised in high yields as the condensation product of aminoferrocene **9** and acetylacetone. Coordination attempts to a variety of metal precursors were unsuccessful and often lead to oxidation of the ferrocene moiety. Also, the steric demand of the ferrocenyl substituents appears to be too large, leaving no free space for a metal fragment.

While optimizing the synthesis of ferrocenylamines, ferrocenylureas were isolated. In preliminary experiments, this novel class of organocatalysts showed promising activities and stereoselectivities in the allylation of acylhydrazones.

# Chapter 3

# Phosphinoferrocenes

The synthesis of novel phosphinoferrocenes is presented, followed by a discussion of the syntheses and applications of the corresponding $Au^I$ and $Ru^{II}$ complexes.

Experimental work was partially carried out by Lukas Sigrist ($Au^I$-chemistry)[139] and Elia Schneider ($Ru^{II}$-chemistry).[140]

## 3.1 Introduction

### 3.1.1 Monodentate Phosphinoferrocenes in Asymmetric Catalysis

Phosphinoferrocenes have mainly gained their reputation as excellent ligands due to the bidentate ligand classes based on dppf and ppfa scaffolds (see Section 1.1.1). Although far less prominent, monophosphino ferrocenes also have proven worth as ligands in catalysis.[141, 142] Ferrocenyldialkylphosphines have been shown to be far less air-sensitive than the corresponding phosphines, while also showing superior reactivities. Representative examples are the applications of ferrocenylphosphines in Baylis-Hillman reactions[143] and rhodium(I)-catalyzed tandem C–H activation/olefin insertion reactions[144] shown in Tables 3.1 and 3.2.

Table 3.1: Ferrocenylphosphines as catalysts in the Baylis-Hillman reaction.[143]

| entry | catalyst | conversion /[%] | yield /[%] |
|---|---|---|---|
| 1 | PCy$_3$ | 24 | 8 |
| 2 | FcPCy$_2$ | 95 | 74 |
| 3 | FcPEt$_2$ | 100 | 98 |

Table 3.2: Ferrocenylphosphines as ligands in rhodium(I) catalyzed tandem C–H activation/olefin insertion.[144]

| entry | ligand | time /[h] | yield /[%] |
|---|---|---|---|
| 1 | PCy$_3$ | 8 | 48 |
| 2 | FcPPh$_2$ | 4 | 34 |
| 3 | FcPCy$_2$ | 2 | 75 |

For the synthesis of chiral monophosphino ferrocenes, one may resort to the methods for stereoselective *ortho*-functionalization that were presented in Section 1.1.1. An alternative approach is the synthesis of *P*-stereogenic ligands. A convenient route towards *P*-stereogenic phosphines is Jugé's ephedrine-based method[145] which has been applied for the synthesis of mono-[146] and diphosphino ferrocenes[147, 148] (Scheme 46). Both planar- and *P*-stereogenic phosphinoferrocenes have found applications in asymmetric catalysis.[141, 142] Representative examples are shown in Schemes 47 and 48.

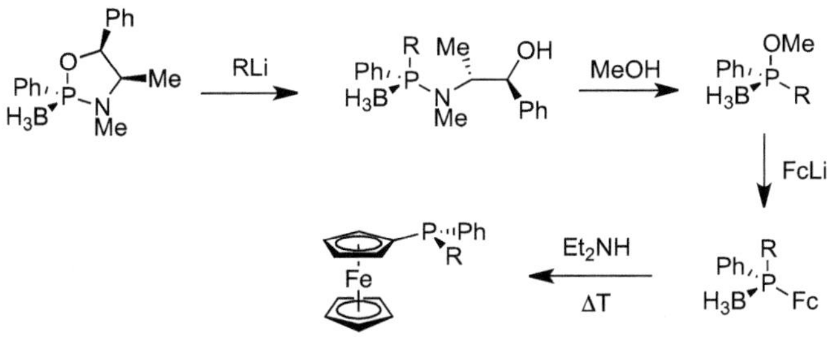

**Scheme 46:** Preparation of *P*-stereogenic monophosphino ferrocenes according to Jamison and co-workers.[146]

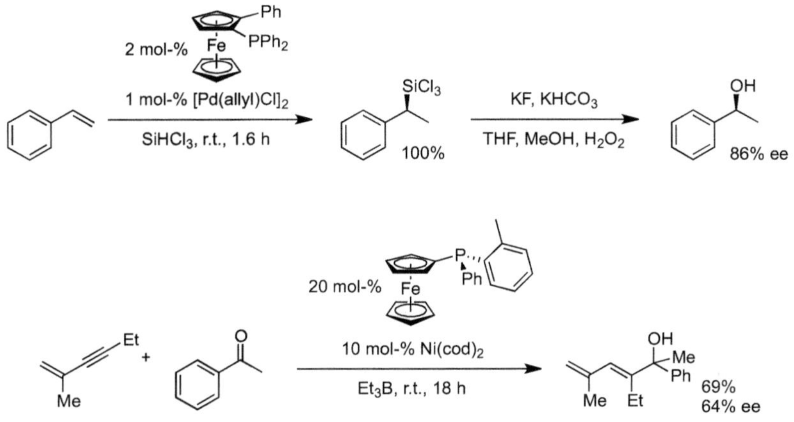

**Scheme 47:** Application of chiral monophosphinoferrocenes: palladium(II)-catalyzed hydrosilylation of styrene[149] (top), nickel(0)-catalyzed reductive coupling of 1,3-enynes and ketones[150] (bottom).

## 3.1.2 Project Idea

As coordination attempts using the prepared aminoferrocenes were not successful (Section 2.3.3), it was decided to prepare the corresponding monodentate phosphinoferrocenes, which are more likely to coordinate to late transition metals according to HSAB theory.[153, 154] Such ligands should be directly accessible starting from the previously

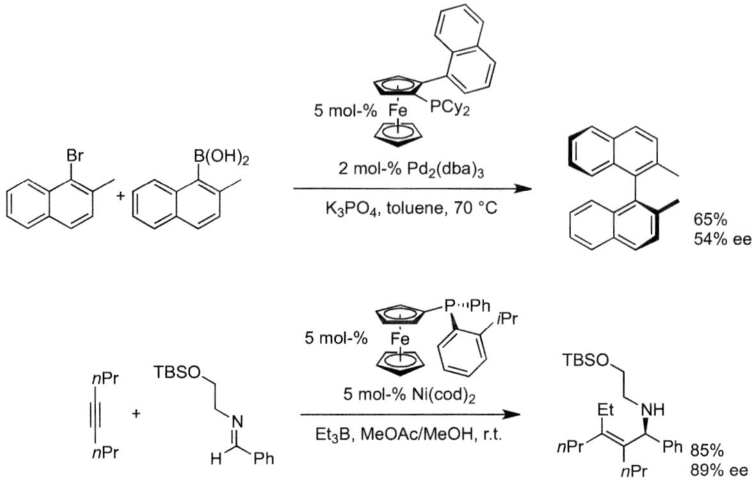

**Scheme 48:** Application of chiral monophosphinoferrocenes: Suzuki cross-coupling[151] (top), nickel-catalyzed three-component coupling of alkynes, imines and organoboranes[152] (bottom).

prepared halides **14–21**. Just as in the initial project idea, the aryl substituent on the side chain of the ferrocene should allow additional arene–metal interactions; at best, a ferrocenyl-tethered complex might be formed (Scheme 49).

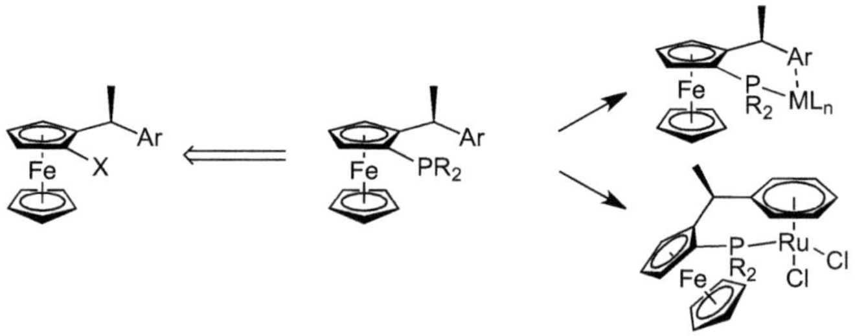

**Scheme 49:** Phosphinoferrocenes should be easily accessible and may act as polydentate ligands.

## 3.2 Ligand Synthesis

Phosphinoferrocenes are usually prepared *via* lithiation of ferrocenes and consequent addition of chlorophosphines. Countless different phosphines are available, which allows both electronic and steric finetuning.

In our case, introducing phosphines should be quite simple. With the bromide **14** in hands, the common procedure for the synthesis of ppfa and its derivatives was followed. Even for bulky phosphines, such as di-*tert*-butyl phosphine, good yields had been reported (Scheme 50).

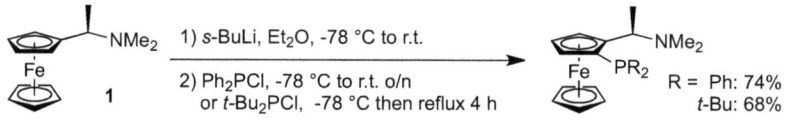

**Scheme 50:** Preparation of ppfa[83] and (*t*-Bu)pfa.[155]

With the presence of a halide, the problem of diastereoselective *ortho*-lithiation had already been addressed. As shown for the preparation of the carboxylic acid **7** in Section 2.3.1.2, transmetalation using *n*-butyllithium gives quantitative conversion. It was therefore assumed that quenching the lithiated ferrocene at low temperature with chlorophosphines and slowly warming up to room temperature should readily yield the desired product. If necessary, the temperature could be raised to reflux for a couple of hours.

However, carrying out the synthesis in diethyl ether using $PClPh_2$ resulted in yields less than 10%, even when the mixture was heated to reflux over night. In order to reach higher temperatures, the solvent was exchanged to THF, even though it is known that in this solvent the lifetime of organolithium compounds may drop drastically.* While keeping all other conditions the same, a yield of 35% was reached. The reaction was monitored by $^{31}P\{^1H\}$ NMR in order

---

*e. g. half-life time of *n*-BuLi: $t_{1/2}$ ($Et_2O$, 35 °C) = 31 h; $t_{1/2}$ (THF, 35 °C) = 10 min.[156]

to keep track of the phosphorus species. In addition to the product signal and small amounts of oxidized phosphine, a substantial amount of tetraphenyldiphosphine was detected. Therefore, an excess of diphenylchlorophosphine had to be used; best results were reached when three equivalents of chlorophosphine were added and the mixture was stirred at reflux over night. Using these conditions, a range of ligands was prepared; yields are given in Table 3.3.

Table 3.3: Preparation of monophosphinoferrocenes.

| entry | bromide | Ar | R | product | yield /[%] |
|---|---|---|---|---|---|
| 1 | 14 | Ph | Ph | 40 | 67 |
| 2 | 14 | Ph | t-Bu | 41 | - |
| 3 | 14 | Ph | i-Pr | 42 | 76[a] |
| 4 | 14 | Ph | Cy | 43 | 80[a] |
| 5 | 16 | 3,5-xylyl | Ph | 44 | 71 |
| 6 | 17 | 2,4,6-mesityl | Ph | 45 | 58 |
| 7 | 18 | 3,5-di-tert-butylphenyl | Ph | 46 | 68 |
| 8 | 19 | 3,5-bis(trifluoromethyl)phenyl | Ph | 47 | -[b] |
| 9 | 20 | 1-naphthyl | Ph | 48 | 70 |
| 10 | 21 | 8-fluoro-1-naphthyl | Ph | 49 | -[b] |

[a]Purification via $BH_3$ protection. [b]Fluoride abstraction was observed.

When diphenyl(ferrocenyl)phosphines were prepared, yields of about 70% were reached in most cases. Only in case of the mesityl substitued bromide **17** did the yield drop below 60%, which may be traced back to the steric influence of the methyl groups in *ortho*-position. No product formation could be observed for fluorinated compounds **19** and **21**. While lithiation occurred selectively on the ferrocene, $^{19}F$ NMR analysis of the crude products indicated fluoride abstraction. In case of the fluoronaphthyl derivative **21**, no signal could be detected at all in a $^{19}F$ NMR analysis of the product mixture. Besides this intolerance towards additional halides, arylphosphines could now be prepared in a reproducible fashion. All products are bench-stable solids; four of them could be characterized by X-ray diffraction (*vide infra*).

Dialkylphosphines, however, were much more difficult to prepare. Due to their sensitivity, borane protection was needed in order to purify the crude products. Even though the protected phosphines were stable enough for column chromatography on silica gel, they soon started to decompose, even when stored under argon, and could not be characterized thoroughly. Nevertheless, good yields could be reached after deprotection with morpholine for both di-*iso*-propyl- and dicyclohexylphosphine. Di-*tert*-butylchlorophosphine, however, could not be introduced; in several attempts, $^{31}$P{$^{1}$H} NMR analysis could not confirm product formation. Presumably, its steric demand is simply too large.

## 3.2.1 Crystal Structures

Aside from the mesityl-substituted ligand **45**, which forms fibrous, microcrystalline needles, all diaryl phosphines could easily be recrystallized to yield X-ray quality crystals. Due to their sensitivity, crystallization of the dialkyl phosphines **42** and **43** was not attempted. ORTEPIII representations of ligands **40**, **44**, **46** and **48** are given in Figure 17 and selected bond and torsion angles are given in Table 3.4.

Overall, the conformation of the free ligands is as expected. Although the structural overlay presented in Figure 18 shows some differences in the orientation of the sidechain, they can largely be traced to intermolecular edge to face $H - \pi$ interactions.[157, 158] Interestingly, only the naphthyl substituted ligand **48** shows weak intramolecular $\pi - \pi$ stacking interactions. In this case, the *exo*-phenyl group of the phosphine and the naphthyl group on the sidechain are aligned at an interplanar distance of 3.6–3.8 Å. The differing orientations of the phosphine groups can also be traced back to the mentioned interactions. The sum of the angles around the phosphorus atoms range from 302.4(6)° (**44**) to 304.6(2)° (**48**); although the difference is statistically not significant, the angles are very similar.

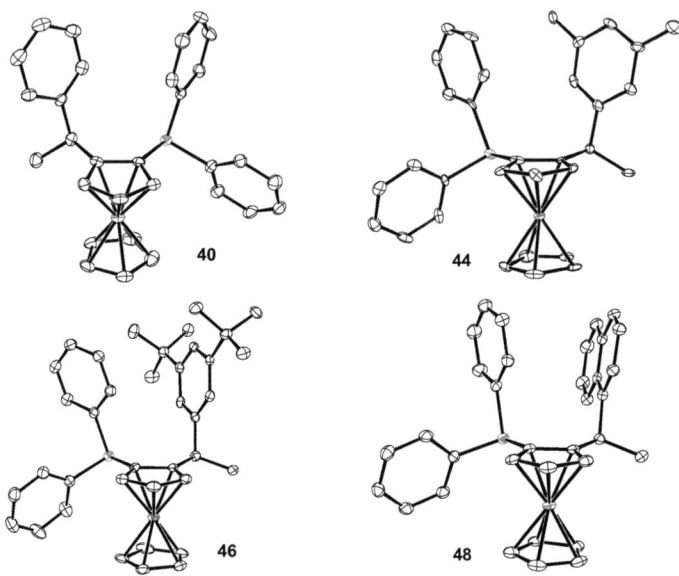

**Figure 17:** ORTEPIII representations of ligands **40**, **44**, **46** and **48**. Hydrogen atoms are omitted for clarity, thermal ellipsoids are set to 50% probability. Selected bond and torsion angles are given in Table 3.4. The asymmetric unit for **44** contains two molecules, an overlay is given in Appendix 6.1.14.

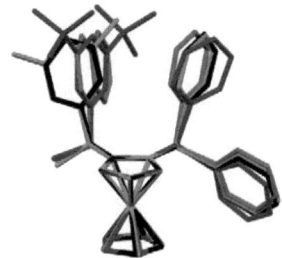

**Figure 18:** MERCURY-generated structural overlay of phosphines **40** (black), **44** (green), **46** (blue), and **48** (red). When neccessary, structures were inverted to show (*S*,*S*)-derivatives.

**Table 3.4:** Selected bond and torsion angles of compounds **40**, **44**, **46** and **48**.[a]

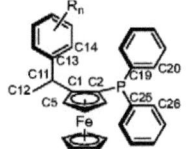

|  | 40 | 44[b] | 46 | 48 |
|---|---|---|---|---|
| C1–C11–C12 /[°] | 114.2(3) | 112.0(8) / 113.6(7) | 112.38(19) | 112.2(2) |
| C1–C11–C13 /[°] | 109.3(3) | 111.7(8) / 110.1(6) | 109.99(18) | 111.1(2) |
| C12–C11–C13 /[°] | 111.2(3) | 111.1(7) / 111.5(7) | 110.85(19) | 108.8(2) |
| C5–C1–C11–C12 /[°] | −45.5(5) | −45.6(12) / −41.9(10) | −23.3(3) | −23.7(3) |
| C5–C1–C11–C13 /[°] | 79.8(4) | 79.8(11) / 84.0(9) | 100.8(3) | 98.3(3) |
| C1–C11–C13–C14 /[°] | 95.5(4) | 92.0(10) / 87.5(9) | 114.9(2) | 132.8(2) |
| C2–P1–C19 /[°] | 102.39(15) | 101.1(4) / 103.8(4) | 100.95(10) | 102.47(11) |
| C2–P1–C25 /[°] | 101.40(15) | 101.4(4) / 99.4(4) | 101.30(10) | 100.40(12) |
| C19–P1–C25 /[°] | 100.82(15) | 99.9(4) / 99.5(4) | 100.66(11) | 101.78(11) |
| C1–C2–P1–C19 /[°] | 98.8(3) | 99.8(7) / 103.7(7) | 103.8(2) | 92.1(2) |
| C1–C2–P1–C25 /[°] | −157.3(3) | −157.6(7) / −154.1(7) | −152.84(19) | −163.21(19) |
| C2–P1–C19–C20 /[°] | −175.4(3) | −171.9(7) / −175.3(7) | −162.21(18) | −176.4(2) |
| C2–P1–C25–C26 /[°] | −73.3(3) | −75.0(8) / −72.1(7) | −76.1(2) | −76.6(2) |

[a] Angles normalized for (*S*,*S*)-derivatives. [b] The asymmetric unit contains two molecules.

# 3.3 Gold(I)-Complexes in Asymmetric Catalysis

## 3.3.1 Introduction

While the first reports of homogenous gold(I) catalysis date back to the early 1930s[159] and the first enantioselective catalysis was presented by Hayashi *et al.* in 1986,[160] the field has received much more attention throughout the last decade. Nowadays, after a veritable gold rush, gold catalysts find countless applications throughout synthetic chemistry,[161, 162] focussing on the electrophilic activation of $\pi$-systems.[163–165] This short intoduction discusses the concepts applied in stereoselective gold(I) catalysis.[166–168]

**Substrate-Controlled Chirality Transfer**

Gold(I) complexes have been shown to be powerful catalysts for reactions involving $\pi$-activation of C–C multiple bonds, such as cycloisomerization[169] and hydroamination [170] of allenes and alkenes, and rearrangements of propargyl ethers.[171, 172] Representative examples are shown in Scheme 51.

Clearly, these enantioselective transformations were catalyzed by non-chiral complexes. The stereoinformation is already present in the chiral starting materials and transferred to the products. Nevertheless, such reactions are often referred to as asymmetric gold catalyses.

**Ligand-Controlled Asymmetric Catalysis**

The first asymmetric gold(I)-catalyzed reaction is of the above-mentioned type. Hayashi and co-workers reported the preparation of optically active oxazoline derivatives *via* a gold(I)-catalyzed asymmetric aldol reaction (Scheme 52).[160] This reaction represents the first example of a catalytic asymmetric aldol reaction and has been investigated thoroughly.[161]

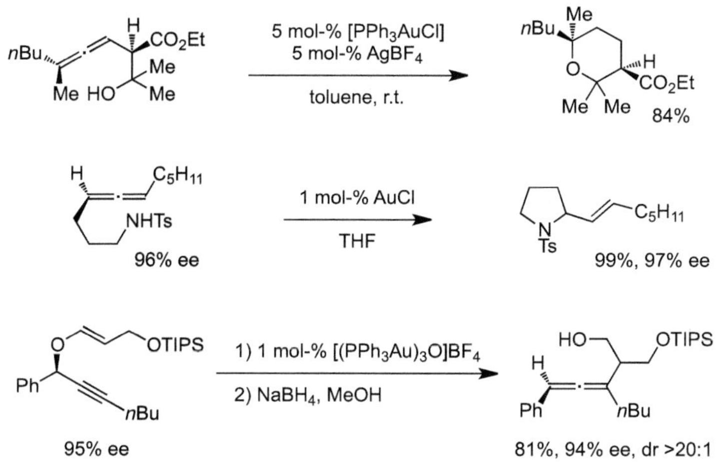

**Scheme 51:** Representative substrate-controlled catalyses by Krause,[169] Yamamoto,[170] and Toste[171] (top to bottom).

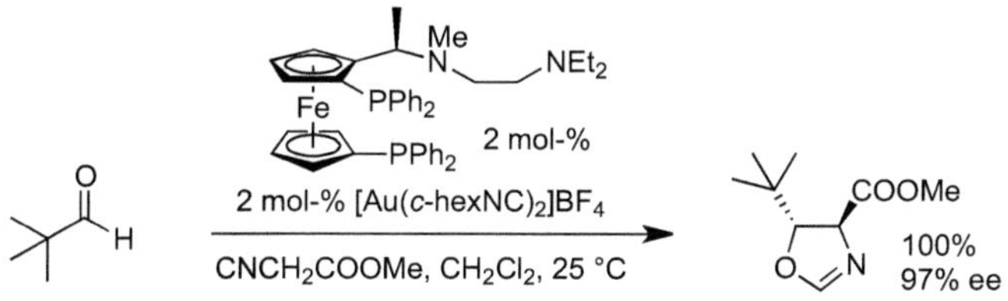

**Scheme 52:** Gold(I)-catalyzed reaction of aldehydes with isocyanoacetate.

As indicated in Figure 19, various coordination geometries of gold(I) are known. However, the chemistry of gold(I)complexes is dominated by linear two-coordinate species. As a consequence, the reacting substrate is forced to approach the reactive gold center *trans* to the ligand, which hinders the transfer of chiral information.

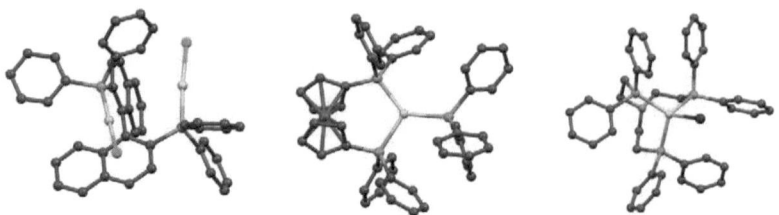

**Figure 19:** Coordination modes in gold(I) complexes (MERCURY representations). Linear coordination in dinuclear [((R)-binap)(AuCl)$_2$][173] (left), trigonal [(dppf)AuPPh$_3$]ClO$_4$ [174] (middle, anion omitted) and tetraedric [(N(CH$_2$CH$_2$PPh$_2$)$_3$AuI][175] (right).

Ligands based on a biaryl scaffold have successfully been applied in the activation of alkenes, allenes and alkynes.[167, 168] Although these are bidentate ligands, the catalysts are usually formed with two equivalents of gold(I) precursors, thereby forming dinuclear complexes with two separated active sites (compare Figure 19, left). An application of a biaryl-based ligand is shown in Scheme 53.

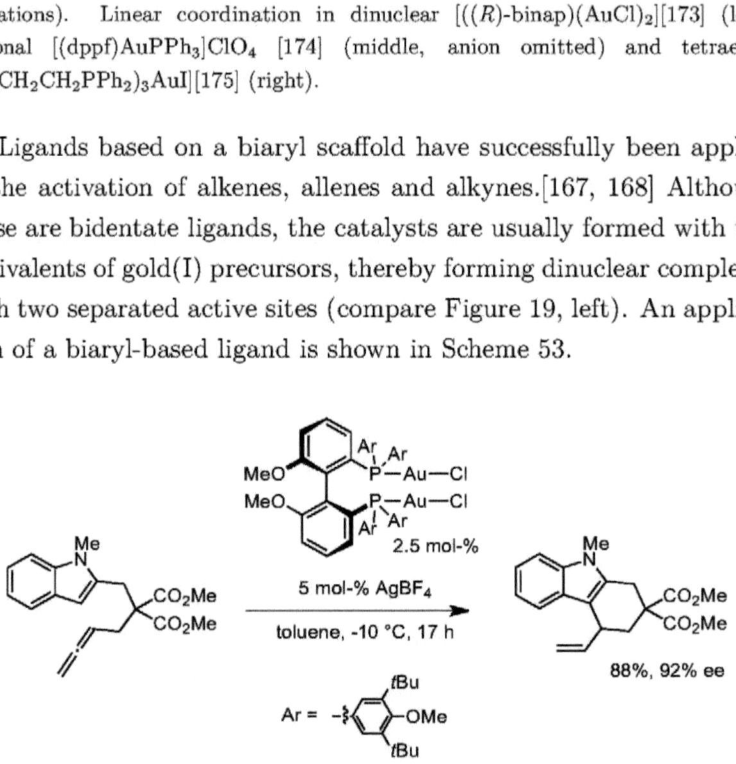

**Scheme 53:** Enantioselective hydroarylation of allenes using a dinuclear biaryl-based ligand.[176]

As observed in the crystal structure of [((R)-binap)(AuCl)$_2$] (Figure 19, left), the biaryl backbone allows the formation of chiral pockets around the active sites. Similarly, monodentate ligands need sterically encumbering substituents in order to induce stereoselectivity. Recent uses of such mononuclear complexes include the application of bulky taddol-based phosphoramidite ligands[177] (Scheme 54) and N-heterocyclic carbenes[178] (Scheme 55).

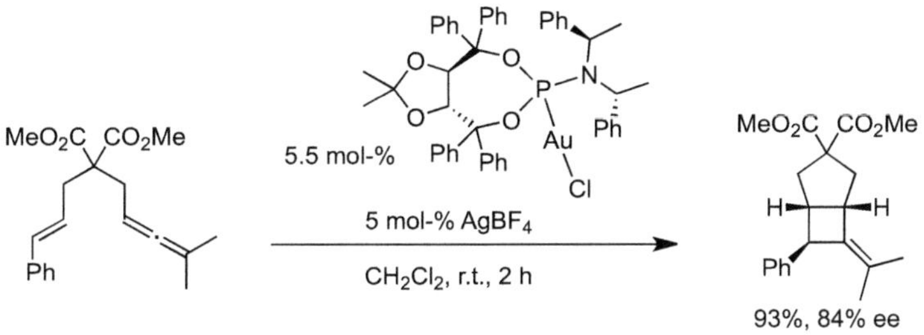

**Scheme 54:** Gold(I)-catalyzed asymmetric [2+2] cycloaddition using taddol-based phosphoramidite ligands.[177]

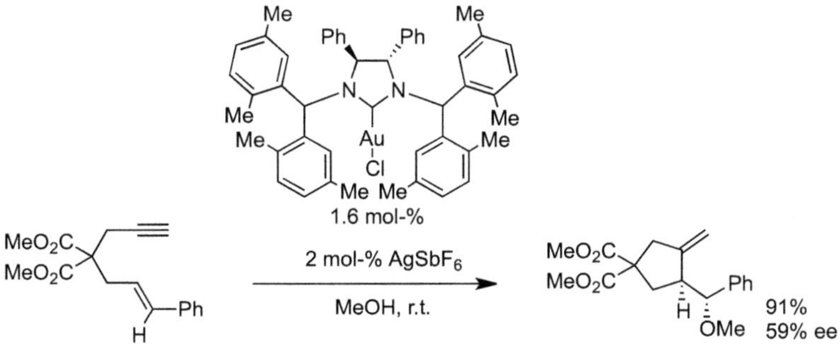

**Scheme 55:** Cyclization of 1,6-enynes catalyzed by gold(I) carbenes.[178]

## Anion Control

Another widely used concept is anion controlled asymmetric catalysis. As the catalyst precursors usually have to be activated by halide abstraction, silver(I) salts with chiral counterions allow the *in-situ* formation of chiral catalysts starting from readily available non-chiral precursors. Of course, this strategy may also be used to enhance stereoinduction of chiral catalyst systems. A new *caveat* in this case is the generation of matched/mismatched situations between ligand and counterion, as encountered by Toste and co-workers (Table 3.5).[179, 180]

Table 3.5: Toste's chiral counterion strategy gives rise to matched/mismatched situations.[179]

| entry | catalyst | yield /[%] | ee /[%] |
|---|---|---|---|
| 1 | [(R)-binap(AuCl)$_2$] | 88 | 83 |
| 2 | [(S)-binap(AuCl)$_2$] | 91 | 3 |

### 3.3.2 Project Idea

Our interest in gold(I) catalysis can be traced back to the initial project idea of a tethered ligand. While the steric bulk of the ferrocenylphosphines **40–48** may already induce stereoselectivity, additional arene–metal interactions would of course lead to a more rigid geometry and possibly more robust catalysts. In the solid state, such interactions have been shown repeatedly for gold(I) complexes of Buchwald-type biaryl phosphines.[181–186] Figure 20 shows two representative complexes with gold(I)–$\eta^2$-arene interactions. The metal–

carbon distances are in the range of 3.06–3.27 Å.

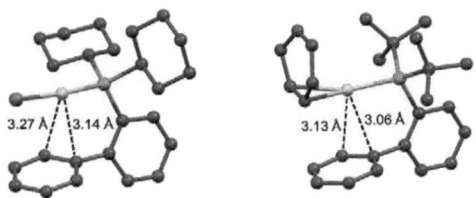

**Figure 20:** MERCURY representations of two complexes showing Au–$\eta^2$-arene interactions with Buchwald-type biaryl phosphines. Left: [AuBr(PCy$_2$(o-biphenyl)];[181] right: [Au(P(t-Bu)$_2$(o-biphenyl)((1,2-$\eta$)-1,4-cyclohexadiene)]SbF$_6$[184] (anion omitted).

In solution, gold(I)–arene $\pi$-interactions remain elusive and have not been confirmed so far. Nevertheless, biarylphosphines often outperform their monoaryl counterparts; an example is given in Table 3.6.[187] On the other hand, calorimetric experiments by Nolan and co-workers on the neutral catalyst precursors showed no sign of a stabilizing interaction in solution.[188]

**Table 3.6:** Gold(I)-catalyzed cycloisomerization of propargyl amides.[187]

| entry | catalyst | t /[h] | yield /[%] |
|---|---|---|---|
| 1 | 2 mol-% **50** | 5 | 87 |
| 2 | [Au(PPh$_3$)]NTf$_2$ (5 mol-%) | 24 | 67 |

In contrast to the rigid conformation of diaryl phosphines, where the pendant aryl ring is forced *per se* into proximity to the metal, the conformation of the aryl groups of our ferrocenylphosphines is far less constrained and preferentially oriented away from the coordination site (Scheme 56). As a result, the observation of metal–arene interactions in the solid state would already indicate a strong stablizing influence, while its absence would not rule out interactions in solution.

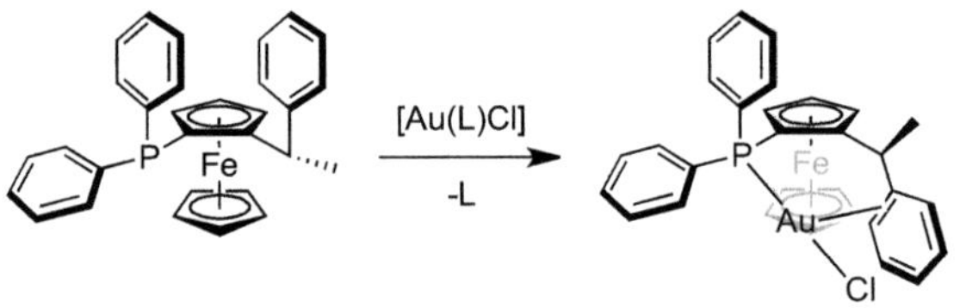

**Scheme 56:** Only a substantial conformational change would allow Au$^I$-arene interactions.

### 3.3.3 Complex Synthesis

(Dimethylsulfide)gold(I) chloride proved to be a convenient precursor for complex synthesis. Coordination of both the phenyl- and the naphthyl-substituted ligands **40** and **48** could be carried out in commercially available dichloromethane, giving the corresponding gold-complexes **51** and **52** in quantitative yields. Crystals suited for X-ray diffraction could be obtained after evaporation of the solvent, while analytically pure product was obtained after removing the solvent incorporated in the crystals *in vacuo*. When technical grade solvent was used, a small drop in yield was observed. In these cases, small amounts of a green precipitate could be observed, indicating the formation of ferrocenium salts,[9] possibly due to the presence of hydrochloric acid in the lower grade solvent.

Even though distilled solvents were used for the coordination of the much more sensitive dialkylphosphines **42** and **43**, a large portion of the ligand underwent oxidation. Because of the inactivity of complex **51** in catalysis (*vide supra*) and a shift of focus towards ruthenium(II) complexes (*vide infra*), these compounds were not examined in detail.

Table 3.7: Synthesis of gold(I)-complexes.

| entry | Ar        | R    | product | yield /[%] | comment             |
|-------|-----------|------|---------|------------|---------------------|
| 1     | Ph        | Ph   | 51      | quant      | p.a. grade solvent  |
| 2     | Ph        | Ph   | 51      | 81         | tech. grade solvent |
| 3     | 1-naphthyl| Ph   | 52      | quant      |                     |
| 4     | Ph        | i-Pr | 53      | n.d.       | Au mirror           |
| 5     | Ph        | Cy   | 54      | n.d.       | Au mirror           |

### 3.3.3.1 Structural Features

The crystal structures of the complexes isolated **51** and **52** are shown in Figure 21. The striking similarities are emphasized in a structural overlay; the geometries are nearly identical.

When comparing the complex structures with the free ligands, it can be seen that instead of a metal–arene interaction, the aryl substituents of the phosphine and the sidechain are much better aligned, allowing strong $\pi-\pi$ stacking interactions (Figure 22). The interplanar distance between the phenyl groups in the all-phenyl complex **51** is 3.5–3.9 Å, while the free ligand showed a weak H–$\pi$ interaction at best. The naphthyl-substituted ligand **48**, on the other hand, already showed $\pi$-stacking. In the complex, the interaction is more pronounced; the phenyl–naphthyl distance is now even shorter at 3.5–3.6 Å and the substituents are better aligned.

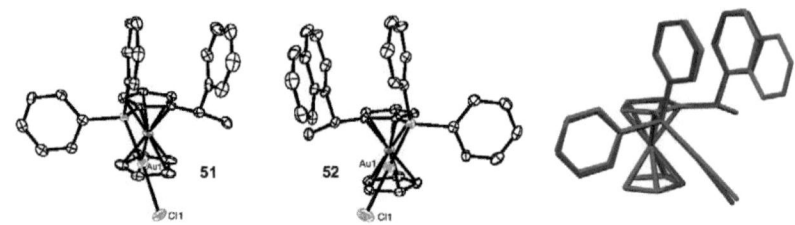

**Figure 21:** ORTEPIII representations of gold(I) complexes **51** (left) and **52** (middle). Hydrogen atoms are omitted for clarity, thermal ellipsoids are set to 50% probability. MERCURY-generated structural overlay (right). Atom numbering similar to Table 3.4. Selected bond lengths [Å], bond and torsion angles [°]: **51**: Au1–Cl1 2.276(2), Au1–P1 2.2256(15), P1–Au1–Cl1 174.31(6), C2–P1–Au1 113.23(19), C19–P1–Au1 116.3(2), C25–P1–Au1 110.69(19), C1–C11–C12 111.7(5), C1–C11–C13 110.4(5), C12–C11–C13 110.4(5), C5–C1–C11–C12 −29.7(8), C5–C1–C11–C13 94.9(6), C1–C11–C13–C14 143.2(6), C2–P1–C19 104.9(3), C2–P1–C25 103.5(2), C19–P1–C25 106.1(3), C1–C2–P1–C19 78.8(5), C1–C2–P1–C25 −170.1(5), C2–P1–C19–C20 −148.0(5), C2–P1–C25–C26 −91.1(5); **52**: Au1–Cl1 2.2792(13), Au1–P1 2.2311(13), P1–Au1–Cl1 177.88(5), C2–P1–Au1 114.18(16), C19–P1–Au1 114.50(16), C25–P1–Au1 112.68(15), C1–C11–C12 112.8(4), C1–C11–C13 110.8(4), C12–C11–C13 110.0(4), C5–C1–C11–C12 −27.1(7), C5–C1–C11–C13 96.7(5), C1–C11–C13–C14 134.5(5), C2–P1–C19 104.9(2), C2–P1–C25 103.2(2), C19–P1–C25 106.4(2), C1–C2–P1–C19 76.9(4), C1–C2–P1–C25 −171.8(4), C2–P1–C19–C20 −144.5(4), C2–P1–C25–C26 −93.8(4).

Attempts were made to cinfirm these interactions in solution by NOESY-experiments, however, no contacts could be detected.[139] While the absence of signals does not necessarily preclude such an interaction, it still shows that it is not of a very strong nature and metal-arene interactions may be possible.

### 3.3.4 Intramolecular Hydroamination

Although this has not been stated so far, the work of Widenhoefer and co-workers on gold(I)-catalysed intramolecuar hydroamination[189] was the initial motivation for this part of the thesis (Scheme 57). Consequently, complex **51** was first applied in a similar fashion.

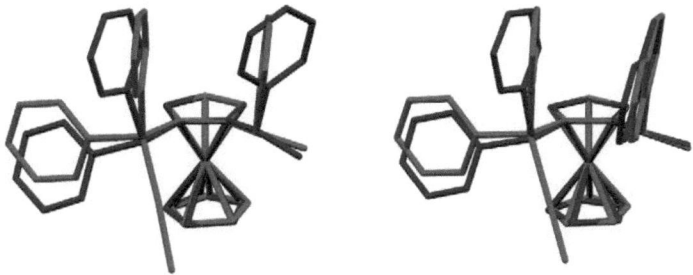

**Figure 22:** MERCURY-generated structural overlays of the complexes prepared **51** (left) and **52** (right) with the corresponding free ligands **40** and **48**.

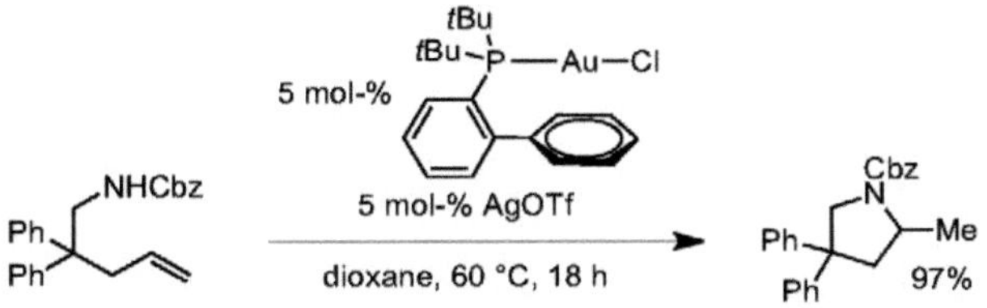

**Scheme 57:** Gold(I)-catalyzed intramolecular hydroamination of alkenes as published by Widenhoefer and co-workers.[189]

The reactions were carried out in freshly purified and degassed dioxane, the catalyst was prepared *in situ* from complex **51** and silver triflate. After one hour, the substrate was added and the reaction mixture was heated to 60 °C and stirred overnight (Scheme 58). Unfortunately, no product formation could be detected, and only a darkening of the mixture was observed. Because a free coordination site is crucial for the formation of an active catalyst, it was decided to first investigate chloride abstraction more thoroughly.

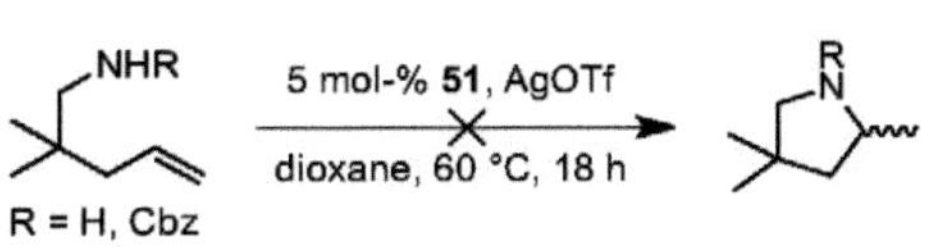

**Scheme 58:** Attemptetd hydroamination using gold(I)-catalyst **51**.

## 3.3.5 Chloride Abstraction

Examining the reaction of silver triflate with catalyst precursor **51** showed that either the silver salt itself oxidizes the ferrocene, even before abstracting the chloride, or that the cationic complex is unstable and decomposes. Therefore a variety of halide scavengers were tested (Table 3.8). Because it had been reported that gold complexes may be sensitive towards ambient light,[190–193] the reactions were carried out in the dark.

Table 3.8: Screening of halogen scavengers.[a]

| entry | AX | solvent | result / comment |
|---|---|---|---|
| 1 | AgOTf | $CD_2Cl_2$ | oxidation of ferrocene |
| 2 | TMSOTf | $CD_2Cl_2$ | no abstraction |
| 3 | $KBF_4$ | $CD_2Cl_2$ | no abstraction |
| 4 | $KPF_6$ | $CD_2Cl_2$ | no abstraction |
| 5 | $NaPF_6$ | $CD_2Cl_2$ | no abstraction |
| 6 | $TlPF_6$ | $CD_2Cl_2$ | no abstraction |
| 7 | $SbCl_5$ | $CD_2Cl_2$ | oxidation of ferrocene |
| 8 | $LiNTf_2$ | $CD_2Cl_2$ | no abstraction |
| 9 | $KBAr_F$ | $CD_2Cl_2$ | $^{31}P$ NMR signal at +32 ppm, <10% |
| 10 | $Et_3OBF_4$ | $CD_2Cl_2$ | $^{31}P$ NMR signal at +32 ppm, <20% |
| 11 | $Et_3OBF_4$ | $CD_3CN$ | $^{31}P$ NMR signal at +32 ppm, <20% |

[a] All reactions were carried out in a *Young*-NMR tube under nitrogen and were monitored by $^{31}P\{^1H\}$ NMR.

Most halide scavengers showed no abstraction. In the cases of silver(I) triflate (entry 1) and antimony(V) pentachloride (entry 7), it could clearly be seen that the ferrocene was oxidized, as the solutions quickly turned blue, the characteristic color of ferrocenium ions.[9] Much to our surprise, most other common scavengers showed no change in the $^{31}$P{$^1$H} NMR spectra. Partially successful chloride abstraction was observed when using KBAr$_F$ (entry 9) ar the Meerwein-salt Et$_3$OBF$_4$ (entry 10). Both showed new signals around +32 ppm in the $^{31}$P{$^1$H} NMR. The best peak-to-peak ratio was reached when using Meerwein's salt representing about 20% abstraction, but such a conversion was still not synthetically useful. In a final attempt, the solvent was exchanged to acetonitrile, which has been shown to stabilize cationic gold(I) species (Figure 23). However, no increase in abstraction was observed.

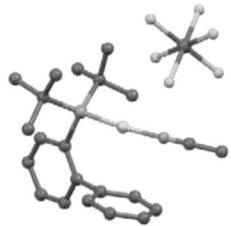

**Figure 23:** Acetonitrile stabilizes the cationic gold(I) species in [Au(P(t-Bu)$_2$(o-biphenyl))(NCCH$_3$)]SbF$_6$ (MERCURY representation).[194]

Due to the poor performance of common halogen scavengers, an alternative approach published by Hintermann *et al.* [195] was tested (Scheme 59). In this method, metal halides and the acid of a weakly coordinating anion react to the hydrogen halides and the corresponding salts; the hydrogen halides are scavenged by epichlorohydrin to give chlorohalidopropanol, which can be removed *in vacuo*. As elegant this approach may be, no product could be isolated.

### 3.3.5.1 An Alkyl Derivative

As the chloride ligand could not be abstracted to any satisfactory degree, an attempt was made to alkylate the gold complex **51** with

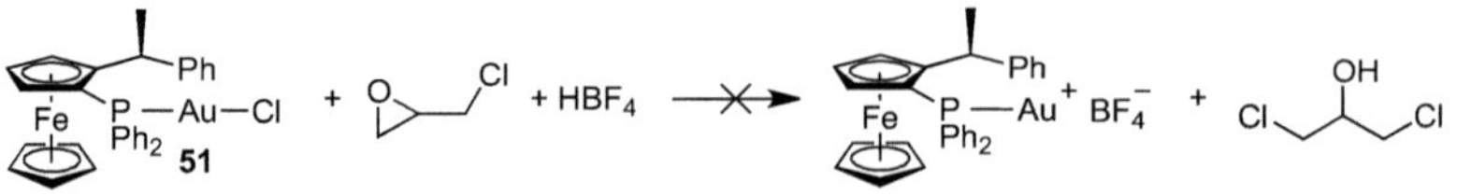

**Scheme 59:** Attempted chloride abstraction using epichlorohydrin according to Hintermann.[195]

methyl magnesium iodide (Scheme 60). Instead of having to scavenge a halide which may stay in solution after all, protonation of the methyl group would lead to the evolution of methane and yield a cationic complex with the conjugate base as counter ion.

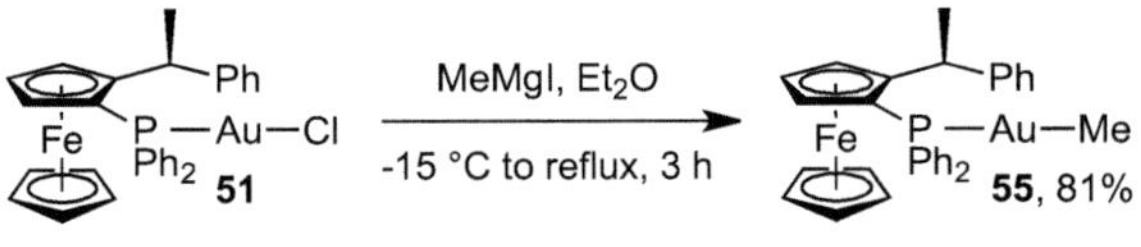

**Scheme 60:** Conversion of chloride **51** to the methyl derivative **55**.

Without further optimization necessary, alkylation of **51** gave full conversion already in the first try. However, purification of the methyl derivative proved very difficult. While chlorinated solvents were best suited in terms of solubility, the product reverted slowly to the chloride, even when freshly distilled chloroform or methylenechloride were used. Therefore, it was not possible to isolate the product in analytically pure form. When measured 5 min after dissolution in $CD_2Cl_2$, $^{31}P\{^1H\}$ NMR already showed approximately 3% reversion to the chloro complex, increasing to 30–40% after 6 h.

**3.3.5.1.1 Crystal Structure** Neglecting its sensitivity towards chlorinated solvents, X-ray quality single crystals were grown from chloroform/pentane. Solving the structure showed that the measured crystal was about a one to one mixture of product **55** and chloro complex **51**. The exact occupancy could not be determined as the positions of the carbon and the chlorine could not be refined independently. The refined Au–Cl/Me distance of 2.186 Å lies nicely between the reported values for $R_3P$–$Au^I$–Me (2.04[196]–2.10 Å [197]) and the Au–Cl bond length of 2.2762(16) Å of the starting material. As a comparison, the bond lenghts of chloro(triphenylphosphine)gold(I) and methyl(triphenylphosphine)gold(I) are 2.279[198]–2.291 Å [199] and 2.065, Å [200] respectively. As it can be seen in Figure 24, the product still has essentially the same conformation as the starting material. The small differences may partially be attributed to the incorporation of different solvent molecules ($CH_2Cl_2$ or $CHCl_3$).

**Figure 24:** MERCURY-generated structural overlay of chloro complex **51** and the mixed Me/Cl structure. As the methyl group and the chlorine atom could not be refined independently, the location of the 'chlorine' is indicated by a dot.

### 3.3.5.2 Cationic Gold(I)-Precursors

Another possible method to avoid the abstraction of a halide from the phosphino complex would be to use halide-free metal precursors, such as bis(dimethylsulfide)- or bis(tetrahydrothiophen)gold(I) salts. The latter have recently been used by Braunstein and co-workers in order to investigate the bonding modes of gold(I) iminolates.[192] Both precursors were prepared by addition of a silver(I) hexafluorophosphate to a solution of the chloro complexes in the corresponding sulfides. Precipitated silver(I) chloride was filtered off and the cationic precursors were crystallized from diethylether.[201]

The cationic precursors were then mixed with one equivalent of ligand **40**. While the tetrahydrothiophene precursor proved to be quite sensitive and decomposed to give a gold mirror, dimethylsulfide was substituted efficiently. $^{31}P\{^1H\}$ NMR analysis of the crude product showed a broad signal at +35.7 ppm as the main product and an impurity at +23.5 ppm which could be be attributed to chloro complex **51**. This comes at no surprise, as the synthesis was carried out in dichloromethane.

Due to the large differences in polarity, the chloro complex could easily be separated from the crude product by column chromatography. The tentative cationic complex was isolated in a yield of 48%. Factoring in the amount of isolated chloro complex (30%), about one fifth of the used phosphine could not be accounted for in the products. This observation was quite puzzling as the reaction showed no signs of decomposition and product loss during purification would have been noticed due to the characteristic coloring of the ligand.

Mass spectroscopic analysis suggested the formation of the cationic bis(ferrocenyl)complex **56**, as shown in Scheme 61. This finding is also consistent with NMR data and the isolated products would account for nearly all the used ligand. Although the yield with respect to the amount of gold(I) precursor drops to 34%, 68% of the ligand was consumed in the formation of this complex.

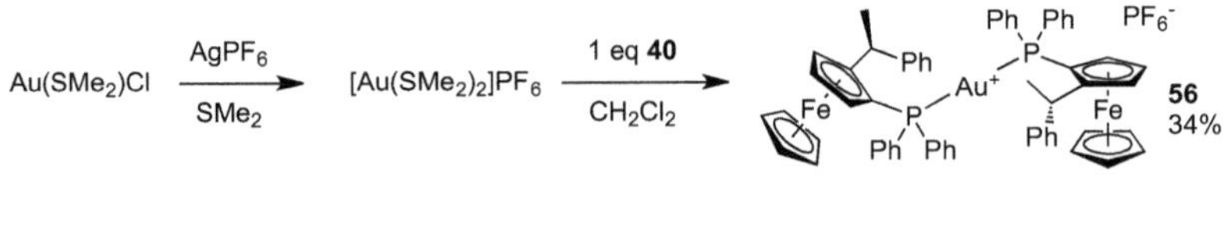

**Scheme 61:** Attempted preparation of cationic complexes lead to the formation of cationic bis(ferrocenylphosphine) complexes.

## 3.3.6 Activation of the Methyl Complex and Hydroarylation of Styrene

Activation of the catalyst was examined before attempting catalysis. As discussed, methyl complex **55** showed high reactivity toward the slightest amounts of hydrogen chloride, hence the main task would not be the activation of the catalyst, but to find an acid which will produce a suitable counter ion, preferentially a non-coordinating one.

In order to stabilize the cationic complexes, activation of **55** was tried in the presence of acetonitrile. The use of acids which should yield common counterions, such as $HBAr_F$, $HBF_4$ or $HNTf_2$, all lead to precipitation of a dark solid, either immediately or within minutes after the addition. Consequently these acids could not be used, especially not in hour-long catalyses. As an exception, activation with sulfuric acid gave a bright red solution and a red precipitate.

Table 3.9: Activation of **55** with acids.[a]

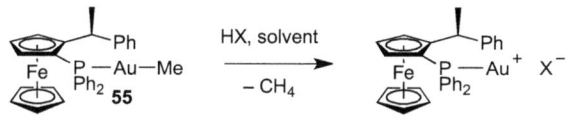

| entry | HX | solvent | result |
|---|---|---|---|
| 1 | $H_2SO_4$ | toluene-$d_8$ | red precipitate |
| 2 | $HBAr_F$ | $CD_2Cl_2$/ $CD_3CN$ 5:1 | dark precipitate |
| 3 | $HBF_4 \cdot Et_2O$ | $CD_2Cl_2$/ $CD_3CN$ 5:1 | slow darkening |
| 4 | $HNTf_2$ | $CD_2Cl_2$/ $CD_3CN$ 5:1 | dark precipitate |

[a] All reactions were carried out in a *Young*-NMR tube.

Looking for a reaction with suitable conditions, hydroarylation of styrene with indoles as published by Che and co-workers [202] seemed applicable, as it was carried out using 2 mol-% [($PPh_3$)AuMe] and 20 mol-% sulfuric acid. In our case, using 5 mol-% **55** and 20–100 mol-% $H_2SO_4$, protonation of the methyl group could be observed (bubbling), but no reaction took place.

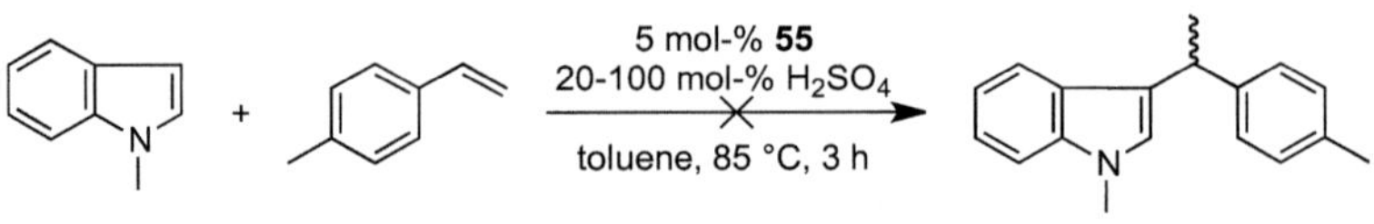

**Scheme 62:** Hydroarylation using gold(I)-catalyst **55** according to Che[202] did not yield product.

# 3.4 Ferrocenyl-Tethered Ru$^{II}$-Complexes for Asymmetric Catalysis

## 3.4.1 Introduction

Over time, the focus of this work had shifted away from the initial idea of synthesizing a bifunctional, tethered catalyst. After the gold(I)-complexes failed to show useful catalytic applications, it was decided to return to the roots of the project and examine whether it was possible to coordinate the phosphine ligands to ruthenium(II). After all, transfer hydrogenation does not strictly depend on bifunctional catalysts, and the applications of ruthenium(II) complexes are not restricted to transfer hydrogenation.[203, 204]

### 3.4.1.1 Transfer Hydrogenation with Phosphine Ligands: Mechanistical Considerations

Transfer hydrogenation catalyzed by transition metals generally operates through hydridic routes (Scheme 63).[48] Although the bifunctional mechanism proposed by Noyori also includes the formation of a metal hydride (see Section 1.2), it stands in strong contrast to non-bifunctional catalysts, where the ligand does not actively take part in the reduction step, and the structure of the transition state is not as well defined. As a consequence, interactions between ligand and substrate are reduced substantially, as indicated in Scheme 64. This should possibly lead to lower stereoselectivities.

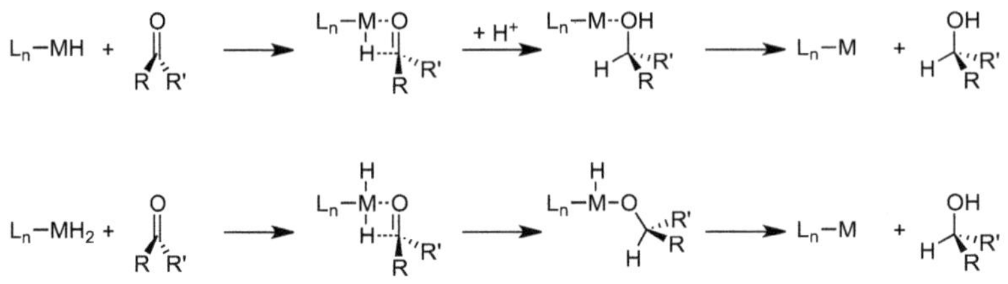

**Scheme 63:** Hydrogenation of ketones *via* mono- (upper) or dihydridic pathways (lower).[48]

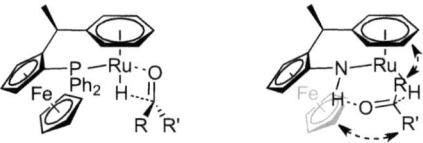

**Scheme 64:** Ligand-substrate interactions are thought to be less pronounced in hydridic pathways (left) compared to a bifunctional mechanism (right).

### 3.4.2 Complex Synthesis

Similar to the aminoferrocene **9**, it was planned to coordinate the phosphine ligands to a suitable ruthenium(II) precursor *via* arene displacement. In initial trials, diphenylphosphine **40** was successfully coordinated to [Ru($\eta^6$-etb)Cl$_2$]$_2$. As the product proved to be stable and the starting material was able to withstand elevated temperatures, the more readily available [Ru($\eta^6$-p-cymene)Cl$_2$]$_2$ could be used as metal precursor. In the following, the reaction conditions found were applied to several of the ligands synthesized earlier in this work (Table 3.10).

Table 3.10: Synthesis of ferrocenyl-tethered ruthenium(II) complexes.

| entry | Ar | R | product | yield | comment |
|---|---|---|---|---|---|
| 1 | Ph | Ph | 57 | 97 | |
| 2 | Ph | $i$-Pr | 58 | 68 | |
| 3 | Ph | Cy | 59 | n.d. | air-sensitive oil |
| 4 | 3,5-xylyl | Ph | 60 | 85 | |
| 5 | 2,4,6-mesityl | Ph | 61 | n.d. | prod. mixture |
| 6 | 3,5-di-*tert*-butylphenyl | Ph | 62 | n.d. | prod. mixture |
| 7 | 1-naphthyl | Ph | 63 | n.d. | prod. mixture |

While the all-phenyl substituted phosphine **40** and the xylyl-ligand **44** gave good to excellent yields, the other arylphosphines gave inseparable product mixtures (entries 5–7). Even after elongated reaction times, not all cymene was displaced by the aryl groups of the ligands. Also, the ligands tend to slowly reduce the ruthenium(II) precursor.

Although it was reasoned that mainly the air-sensitive dialkylphosphines **42** and **43** would be difficult to coordinate and be prone to decomposition, both gave good conversions as long as the ligands were freshly prepared. The di-*iso*-propyl ligand **42** gave the corresponding complex **58** as an air-stable, crystalline solid in 68% yield. The cyclohexyl derivative **43**, however, yielded a brown-red, air-sensitive oil.

With three bench stable complexes in our hands, it was decided to focus on their characterization and catalytic applications.

### 3.4.3 Crystal Structures

X-ray quality single crystals of **57** and **60** were grown *via* diffusion of ethanol into a saturated solution of the complexes, **58** was crystallyzed *via* slow evaporation of a solution of the compound in methanol.

The structure solution of the all-phenyl complex **57** finally proved that it was indeed possible to synthesize ferrocenyl-tethered ruthenium-arene complexes (Figure 25, left). While the orientation of the diphenylphosphino group did not change significantly in comparison to that of the free ligand, the side chain conformation is drastically different. As in the gold(I) complexes, the ruthenium atom is located *endo* with respect to the ferrocene unit. The steric demand of the $RuCl_2$-fragment leads to a distortion of the ligand structure. The two cyclopentadienyl planes are at an angle of 6.8(2)° and the phosphorus atom is located 0.37(2) Å out of the Cp-plane, resulting in a Cp-C2-P1 angle of 11.9(2)°.

The structure of the other all-aryl complex **60** is generally similar to the structure of **57** (see structural overlay in Figure 25). Again, the ruthenium atom is located *endo* with respect to the ligand. The distortion of the ferrocene is less pronounced: the Cp-planes are angled at 4.5(2)° and the phosphorus atom deviates 0.31(2) Å, or 9.5(2)° out of the Cp-plane.

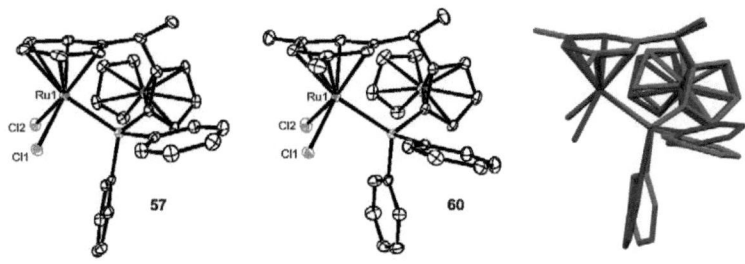

**Figure 25:** ORTEPIII representations of ruthenium(II) complexes **57** (left) and **60** (middle). The asymmetric unit consists of two complexes and one solvent molecule. The second complex, the solvent molecule and all hydrogen atoms are omitted for clarity, thermal ellipsoids are set to 50% probability. MERCURY-generated structural overlay (right). Atom numbering similar to Table 3.4. Selected bond lengths [Å], bond and torsion angles [°]: **57**: Ru1–Cl1 2.4107(7), Ru1–Cl2 2.4060(8), Ru1–P1 2.3171(8), Ru1–arene 1.683(3), P1–Ru1–Cl1 87.92(3), P1–Ru1–Cl2 89.05(3), Cl1–Ru1–Cl2 87.15(3), C2–P1–Ru1 111.90(10), C19–P–Ru1 113.05(10), C25–P1–Ru1 117.76(9), C1–C2–P1–Ru1 −39.2(3), C1–C11–C12 109.0(2), C1–C11–C13 117.6(2), C12–C11–C13 110.7(2), C5–C1–C11–C12 64.6(3), C5–C1–C11–C13 −168.4(3), C1–C11–C3–C14 83.5(3), C2–P1–C19 100.35(13), C2–P1–C25 106.93(13), C19–C1–C25 105.19(13), C1–C2–P1–C19 81.0(3), C1–C2–P1–C25 −169.5(2), C2–P1–C19–C20 −171.6(2), C2–P1–C25–C26 −144.4(2); **60**: Ru1–Cl1 2.4079(6), Ru1–Cl2 2.3890(7), Ru1–P1 2.3213(7), Ru1–arene 1.702(3), P1–Ru1–Cl1 87.98(2), P1–Ru1–Cl2 90.61(3), Cl1–Ru1–Cl2 85.84(2), C2–P1–Ru1 112.03(9), C19–P–Ru1 113.86(9), C25–P1–Ru1 121.28(8), C1–C2–P1–Ru1 −35.7(3), C1–C11–C12 110.3(2), C1–C11–C13 116.7(2), C12–C11–C13 111.4(2), C5–C1–C11–C12 63.1(3), C5–C1–C11–C13 −168.5(2), C1–C11–C3–C14 86.1(3), C2–P1–C19 101.04(12), C2–P1–C25 103.87(12), C19–C1–C25 102.37(12), C1–C2–P1–C19 85.8(2), C1–C2–P1–C25 −168.4(2), C2–P1–C19–C20 −146.7(2), C2–P1–C25–C26 −117.6(2).

As shown in Figure 26, in the dialkylphosphine complex **58**, the ruthenium atom is located *exo* to the ferrocene. This results in a drastic change around the environment of the central atom, as the ferrocene moiety has no direct contact with the metal center. Instead, the coordination sphere around the ruthenium atom is only influenced by the two *iso*-propyl substituents of the phosphine. In contrast to the phenyl groups of the other ligands, one alkyl group is pointing directly in direction of the ferrocene bulk. Again, the ferrocene unit is distorted.[†] The Cp-planes are angled at 9.3(3)°; the phosphorus atoms deviate 0.50(3) Å or 14.8(3)° out of the Cp-plane.

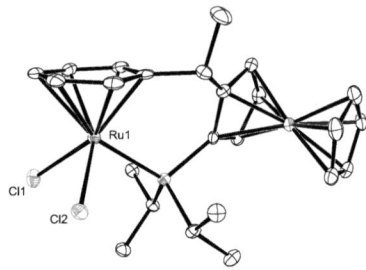

**Figure 26:** ORTEPIII representations of ruthenium(II) complex **58**. The asymmetric unit consists of two complexes and one solvent molecule. The second complex, the solvent molecule and all hydrogen atoms are omitted for clarity, thermal ellipsoids are set to 50% probability. Atom numbering similar to Table 3.4. Selected bond lengths [Å], bond and torsion angles [°]: molecule 1: Ru1–Cl1 2.4330(9), Ru1–Cl2 2.4109(8), Ru1–P1 2.3834(9), Ru1–arene 1.703(3), P1–Ru1–Cl1 96.23(3), P1–Ru1–Cl2 89.02(3), Cl1–Ru1–Cl2 84.18(3), C2–P1–Ru1 111.70(10), C19–P–Ru1 116.97(11), C25–P1–Ru1 115.42(10), C1–C2–P1–Ru1 23.7(3), C1–C11–C12 112.3(3), C1–C11–C13 109.9(3), C12–C11–C13 108.0(3), C5–C1–C11–C12 −1.1(5), C5–C1–C11–C13 119.2(3), C1–C11–C3–C14 148.7(3), C2–P1–C19 101.06(14), C2–P1–C25 106.42(15), C19–C1–C25 103.77(14), C1–C2–P1–C19 148.8(3), C1–C2–P1–C25 −103.1(3); molecule 2: Ru1–Cl1 2.4325(9), Ru1–Cl2 2.4190(8), Ru1–P1 2.3753(9), Ru1–arene 1.694(3), P1–Ru1–Cl1 95.74(3), P1–Ru1–Cl2 89.89(3), Cl1–Ru1–Cl2 85.17(3), C2–P1–Ru1 112.23(10), C19–P–Ru1 116.45(10), C25–P1–Ru1 115.53(11), C1–C2–P1–Ru1 23.3(3), C1–C11–C12 113.0(3), C1–C11–C13 110.1(3), C12–C11–C13 108.8(3), C5–C1–C11–C12 −1.2(5), C5–C1–C11–C13 120.7(4), C1–C11–C3–C14 145.3(3), C2–P1–C19 100.53(14), C2–P1–C25 106.38(14), C19–C1–C25 104.16(14), C1–C2–P1–C19 147.8(3), C1–C2–P1–C25 −103.9(3).

---

[†]The asymmetric unit contains two molecules. If statistically equivalent, mean values with combined standard deviations are given in the text.

### Steric Influence at the Active Site

Although the crystal structures can only hint towards the structure of the active catalysts in solution, one may already assess the steric influence of the ligands. Figure 27 shows overlays of the two $\eta^6$-phenyl ligands from three different perspectives.

The conformation of the ligands with respect to the ruthenium-arene fragment could not be more different (*front* and *top* views in Fig 27). Looking at the structures perpendicular to the metal-phosphorus bond (*side* view), however, reveals that both ligands shield their side quite well but their influence does not reach far into what tentatively may be thought of as the active site.

**Figure 27:** MERCURY-generated structural overlays of **57** and **58**. *Front* (left), *side* (middle), and *top* (right) view.

Overall, the substituents on the all-phenyl ligand cover a larger area and appear to be more constrained, than the ligand in complex **58**. Here, the coordination sphere around ruthenium is mainly influenced by the two isopropyl groups.

### 3.4.4 Asymmetric Transfer Hydrogenation

The three isolated, crystalline complexes **57**, **58** and **60** already allowed for the evaluation of the influence of varying substituents at both the arene and the phosphine.

A general procedure for the catalyses and a table containing additional results are given in Section 5.3.5 in the experimental part.

### 3.4.4.1 Preliminary Screening

It was first tried to get a rough estimate on the catalyst's reactivity in the reduction of acetophenone using the all-phenyl complex **57**; the results are shown in Table 3.11. First attempts were carried out using 5 mol-% catalyst, 5mol-% $Et_3OPF_6$, and 5 mol-% potassium *tert*-butoxide as a base. Full conversion was reached within 1 h at 60 °C, but racemic product was formed (entry 1). This was also the case when the catalyst loading and/or the temperature was lowered, or no chloride was abstracted (entries 2–5).

Examining the effects of the potassium salt showed drastic drops in reactivity when no base was added. After 20 h, 15% conversion was reached (entry 6). Even slower reaction rates were observed when neither base nor Meerwein salt were added (entry 7). The use of triethylamine as non-ionic base led to high yields after 20 h, but still racemic product (entry 8). First enantioselectivites were obtained when triethylamine was used and the chloride was abstracted (entry 9). The use of triethylamine/formic acid as hydrogen donor, gave higher conversions but led again to racemic product.

Table 3.11: Preliminary screening using **57**.

| entry | solvent | base | X⁻ | T/[°C] | tim /[h] | yield/[%] | er |
|---|---|---|---|---|---|---|---|
| 1 | iPrOH | KOt-Bu | PF$_6$ | 60 | 1 | >99 | rac |
| 2 | iPrOH | KOt-Bu | - | 60 | 1 | >99 | rac |
| 3[a] | iPrOH | KOt-Bu | PF$_6$ | 60 | 1 | >99 | rac |
| 4[a] | iPrOH | KOt-Bu | - | 60 | 1 | >99 | rac |
| 5[a] | iPrOH | KOt-Bu | - | 40 | 2 | >99 | rac |
| 6 | iPrOH | - | PF$_6$ | 60 | 20 | 15 | rac |
| 7 | iPrOH | - | - | 60 | 20 | <10 | rac |
| 8 | iPrOH | NEt$_3$ | - | 60 | 20 | 94 | rac |
| 9 | iPrOH | NEt$_3$ | PF$_6$ | 80 | 20 | 25 | 62:38 |
| 10 | NEt$_3$/HCOOH[b] | - | PF$_6$ | 80 | 20 | 62 | rac |

[a] substrate:catalyst:X⁻:base 100:1:1:1. [b] NEt$_3$/HCOOH 2:5.

In summary, the turnover frequency of the catalyst is quite low, probably around a couple of turnovers per hour. More importantly, it cannot keep up with with other cations in solution, so no salts should be used as bases and halide scavengers should not contain hard cations. On the other hand, nitrogen bases helped to accelerate the reaction. Abstracting one chloride and using a base finally reduced the rate of the reaction, but resulted in optically active product.

In the following, the influence of different bases and halide scavengers are examined, using the conditions described in entry 9, Table 3.11, as standard conditions.

### 3.4.4.2 Base Screening

A variety of nitrogen bases were tested and the results are summarized in Table 3.12. When no halide scavenger was used, only racemic product was obtained (entries 1, 4, 6 and 7), which is in agreement with the above discussed results. The catalysis using $NEt_3$ could nicely be reproduced (entry 2). Adding more base led to a slight decrease in yield while similar selectivity was obtained. The addition of pyridine led to catalyst poisoning; only 2% conversion was observed (entry 5). A color change of the mixture was also observed upon pyridine addition.

Better results were obtained when using secondary amines. Dimethylamine showed a similar selectivity to triethylamine, but twice the yield was reached. Di-*iso*-propyl amine gave the best result so far, yielding 55% alcohol with an enantiomeric ratio of 68:32.

Table 3.12: Base screening using **57**.

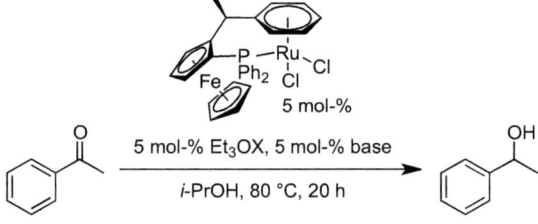

| entry | base | X⁻ | yield [%] | er |
|---|---|---|---|---|
| 1 | $NEt_3$ | - | 95 | rac |
| 2 | $NEt_3$ | $PF_6$ | 25 | 62:38 |
| 3[a] | $NEt_3$ | $PF_6$ | 18 | 61:39 |
| 4 | dbu | - | 93 | rac |
| 5 | dbu | $PF_6$ | 61 | rac |
| 6 | dipea | - | 72 | rac |
| 7 | pyridine | - | 2 | rac |
| 8 | NH*i*-Pr$_2$ | $PF_6$ | 55 | 68:32 |
| 9 | $NHMe_2$ | $PF_6$ | 45 | 62:38 |

[a] 10 mol-% base.

### 3.4.4.3 Halide Abstraction

As a final optimization step before the other complexes were tested, the effect of different counterions was examined. As inorganic cations catalyzed the reaction faster than the ruthenium catalyst itself, it was decided to use Meerwein's salts. As shown in Table 3.13, large counterion effects were observed.

Table 3.13: Counterion effects when using **57**.

| entry | base | X⁻ | yield [%] | er |
|---|---|---|---|---|
| 1 | NEt$_3$ | - | 95 | rac |
| 2 | NEt$_3$ | BF$_4$ | 5 | n.d. |
| 3 | NEt$_3$ | PF$_6$ | 25 | 61:39 |
| 4 | NEt$_3$ | SbF$_6$ | 50 | 67:33 |

As previously observed, the yields of the catalysis dropped substantially when halide scavengers were used, most pronounced when tetrafluoroborate was used (entry 2). In this case only about one turnover was reached. The small amount of product was not sufficient to determine enantioselectivity, and as the conversion was so sluggish, its selectivity was not further examined. Hexafluoroantimonate, on the other hand, gave twice the yield and enhanced selectivity (entry 4) compared to the hexafluorophosphate salt.

Halide abstraction itself was then examined more closely by $^{31}$P{$^1$H} NMR. It was noticed that the tetrafluoroborate reacted much slower than the other two salts. When using one equivalent of scavenger, full conversion was reached after two days. The phosphate reacted much faster, but led to the formation of additional sideproducts. The cleanest and fastest abstraction was observed with the hexafluoroantimonate salt. Full conversion could be reached within half an hour and only one product was formed. This was also observed visually, as the bright red color of the dissolved complex quickly turned into a deep red after adding the scavenger.

The extend of ion pairing in solution could not be determined, as the two scavengers giving clean products showed very broad signals in the corresponding $^{19}$F NMR spectra and the the $PF_6$-salt showed too many side products.

### 3.4.4.4 Ligand Effects and Final Optimization for the Transfer Hydrogenation of Acetophenone

Under the optimized conditions in hand, the three available ligands and the information gained thus far were used to optimize the reduction of acetophenone. The results are shown in Table 3.14.

As hexafluoroantimonate salts were found to give the best results, the effect of different bases was examined first. When using $PF_6^-$ as counter ion, it had been observed that doubling the amount of $NEt_3$ led to a decrease in yield while enantioselectivity stayed the same. This time, the drop in yield was accompanied by a loss of selectivity; racemic product was obtained (entry 2). Also in accordance with previous observations, using dbu increased the yield significantly but gave racemic product (entry 3). For the all-phenyl complex **57**, the best reactivity and selectivity were reached when applying the $SbF_6^-$-salt with di-*iso*-propylamine as the base (entry 4). After 20 h at 80 °C, 60% yield in an enantiomeric ratio of 72:28 could be obtained. Double chloride abstraction led to a negligible increase in yield (entry 5).

The xylyl-complex **60** exhibited an overall lower performance, giving 35% yield with a selectivity of 58:42 (entry 8). Using di-*iso*-propylamine as a base led to slightly improved performance (entry 10), while $PF_6^-$ had a negative effect on the reaction rate and exhibited slightly lower stereoselectivity.

When examining the reactivity of the dialkylphosphino complex **58**, a strong influence of the base was observed (entries 11–14). While triethylamine gave 40% yield, yields were doubled when using dimethylamine. The enantioselectivities were in the range of those of the xylyl-complex **60**, but in favor of the opposite enantiomer, ranging from 46:54 (NEt$_3$, entry 11) to 40:60 (NH$_3$, entry 12).

Table 3.14: Final screening using **57**, **58** and **60**.

| entry | catalyst | base | X$^-$ | yield [%] | er |
|---|---|---|---|---|---|
| 1 | 57 | NEt$_3$ | SbF$_6$ | 50 | 67:33 |
| 2$^a$ | 57 | NEt$_3$ | SbF$_6$ | 40 | rac |
| 3 | 57 | dbu | SbF$_6$ | 81 | rac |
| 4 | 57 | NH$i$-Pr$_2$ | SbF$_6$ | 59 | 72:28 |
| 5$^b$ | 57 | NH$i$-Pr$_2$ | SbF$_6$ | 61 | 72:28 |
| 6 | 57 | NHMe$_2$ | SbF$_6$ | 49 | 70:30 |
| 7 | 57 | NH$_3$ | SbF$_6$ | 51 | 63:37 |
| 8 | 60 | NEt$_3$ | SbF$_6$ | 35 | 58:42 |
| 9 | 60 | NEt$_3$ | PF$_6$ | 23 | 56:44 |
| 10 | 60 | NH$i$-Pr$_2$ | SbF$_6$ | 38 | 61:39 |
| 11 | 58 | NEt$_3$ | SbF$_6$ | 38 | 46:54 |
| 12 | 58 | NH$_3$ | SbF$_6$ | 59 | 40:60 |
| 13 | 58 | NHMe$_2$ | SbF$_6$ | 81 | 42:58 |
| 14 | 58 | NH$i$-Pr$_2$ | SbF$_6$ | 67 | 44:56 |

$^a$10 mol-% base. $^b$10 mol-% halide scavenger.

### 3.4.4.5 Conclusions

The catalyses performed show both base dependence and counter ion effects. Hexafluoroantimonate could easily be determined to be the best counter ion (see Table 3.13). Only minor differences in performance were observed between single and double halide abstraction.

The effects of varying the base are less easily characterizable. Basic salts such as potassium-*tert*-butanolate outperform the transition metal catalyst by themselves, the addition of dbu led to racemic product, and pyridine poisons the catalyst. Amines appear to be well suited, but no trend valid for all catalysts could be observed.

The overall modest selectivities are in accordance with the observations made when examining the crystal structures (Section 3.4.3). In all three cases, the steric influence on the active site appears to be quite small. The inversion of stereoselectivity can be explained by the different conformations of the ligands (Figure 28). While in the case of the arylphosphines the ferrocenyl unit accounts for the largest steric bulk, it does not interact with the active site in case of the di-*iso*-propylphosphine **58**. On the other hand, the conformation of the *iso*-propyl groups favors the opposite selectivity.

**Figure 28:** MERCURY-generated structural overlay of **57** and **58**. The differing steric interactions may account for the inversion of enantioselectivity.

## 3.5 Additional Investigations

### 3.5.1 Coordination to Other Transition Metals

The all-phenyl ligand **40** was also coordinated to other transition metals. Bench-stable and fully characterizable complexes could be obtained in the reactions of **40** with [Rh(Cp*)Cl$_2$]$_2$ and [Ir(Cp*)Cl$_2$]$_2$ in dichloromethane. Both the iridium(III) complex **64** and rhodium(III) complex **65** could be isolated as single crystals, and their solid state structure was determined. The structures are congruent within experimental error; a structural overlay shows only neglectable differences (Figure 29). An interesting feature is that in order to minimize steric interactions, the phosphine group has rotated by about 120° around the Cp–P bond. As a result, the chiral side chain of the ferrocene unit does not interact with the metal center at all but possibly interacts intramolecularly via $\pi$-stacking with the *exo*-phenyl group of the phosphine.

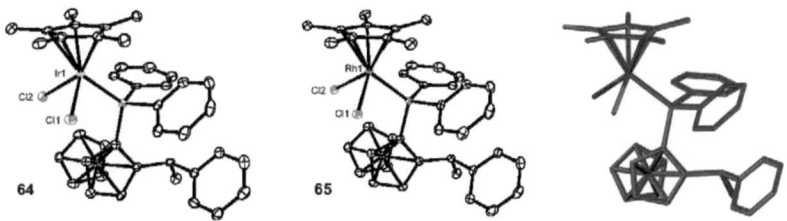

**Figure 29:** ORTEPIII-representation of Ir(III)-complex **64** (left) and Rh(III) complex **65** (middle). Right: MERCURY-generated structural overlay of **64** and **65**.

Rhodium(I) and iridium(I) precursors, on the other hand, gave more sensitive products. Already during the reaction, reduction of the metals was observed. Coordination of **40** to dimethylsulfide copper(I) bromide yielded an air-stable, amorphous solid which was not analytically pure.[139] Platinum(II)- and palladium(II)-precursors gave dynamic mixtures of products.

## 3.5.2 Alternative Ligands

### 3.5.2.1 Indole at the Side Chain

While the Curtius-rearrangement used in the synthesis of the primary aminoferrocene **9** precludes the presence of other functional groups, phosphines are routinely introduced in the presence of directing groups, which can afterwards be nucleophilically substituted. Using this methodology starting from Ugi's amine (**1**), countless ligands are available within two synthetic steps.

As the synthesis of our ligands showed varying yields for the introduction of the aryl groups and introducing the phosphines is only achieved when using large excesses of chlorophosphines, a reversed route could have several advantages. Starting from Ugi's amine, the route would lead *via* ppfa-type phosphines of which numerous examples are known in the literature.[205]

The dimethylamino group of ppfa-type ligands has successfully been substituted with nitrogen-containing heteroarenes such as pyrazoles and indazoles to give P,N-bidentate ligands.[206] In our case, the main goal was to introduce an arene, therefore 1*H*-indole was chosen as nucleophile.

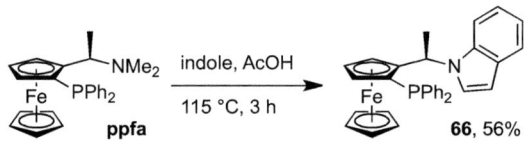

**Scheme 65:** Synthesis of **66** starting from ppfa.

As depicted in Scheme 65, the reaction was carried out successfully in 56% yield. The product could be isolated as orange microcrystals. Single crystals could be grown by slow diffusion of ethanol into a saturated solution of **66** in ether. The crystal structure is shown in Figure 30.

Attempts were made to coordinate this ligand to gold(I) and ruthenium(II) precursors. In both cases, partial reduction of the metal was observed, along with the formation of a complex mixture of products (Scheme 66).

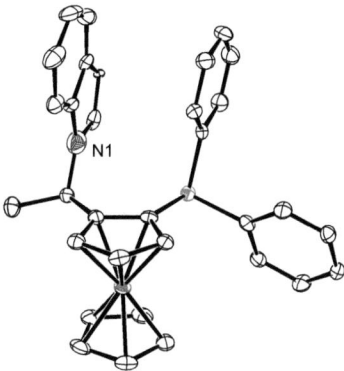

**Figure 30:** ORTEPIII-representation of **66**. The asymmetric unit contains two molecules, only one is shown. Hydrogen atoms are omitted for clarity, thermal ellipsoids are set to 50% probability.

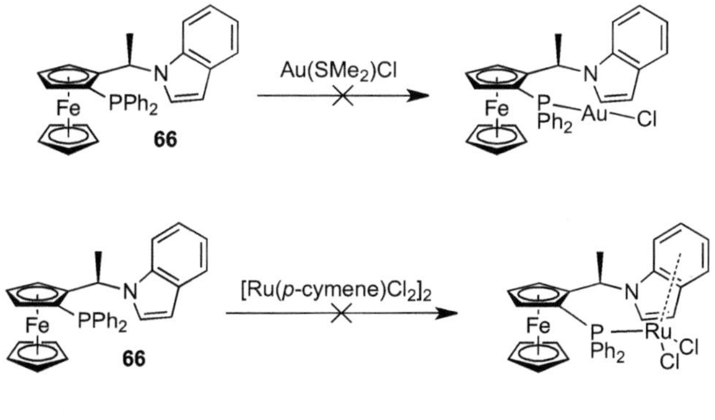

**Scheme 66:** Coordination attempts of **66** to $Au^I$- and $Ru^{II}$-precursors failed.

### 3.5.2.2 Secondary Phosphine Oxides

Secondary phosphine oxides have been shown to act as bifunctional ligands. As the moderate asymmetric induction of our ligand appeared to be caused by a deficiency of influence of the ligand towards the active site, a bifunctional ligand could substantially increase the stereoselectivity of the catalysts (Scheme 67).[207]

**Scheme 67:** Tri- and pentavalent forms of the projected phosphine oxides.

Secondary phosphine oxides can be prepared from less bulky dichloroaryl- or dichloroalkyl phosphines.[208] This could be advantageous in our case, as our ligands needed threefold excess of phosphines and bulky phosphines like chlorodi-*tert*-butyl phosphine did not afford product at all. However, the reaction of bromoferocene **14** with neither *t*-BuPCl$_2$ nor PhPCl$_2$ yielded product (Scheme 68).

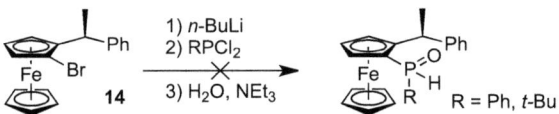

**Scheme 68:** The reaction of bromide **14** with dichlorophosphines yielded no products.

### 3.5.2.3 Sulfoxides

As shown in Section 1.1.1, chiral sulfoxides can be used as directing groups for the synthesis of enantiopure 1,2-disubstituted ferrocenes. In the case of ferrocenyl aryl-sulfoxides, this moiety may coordinate to a metal fragment *via* the sulfur- or oxygen atom or the aromatic system. As our ligands tended to be oxidized by metal fragments, especially in the case of gold(I) complexes **51** and **52**, additional donor atoms could stabilize cationic complexes. However, sulfoxide **67**[‡] efficiently reduced gold(I)- and ruthenium(II)-precursors (Scheme 69).

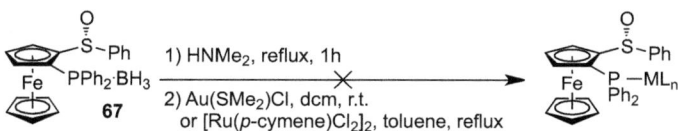

**Scheme 69:** Sulfoxide **67** could not be coordinated to $Au^I$- or $Ru^{II}$-precursors.

### 3.5.3 Attempted Trifluoromethylation of $\beta$-Keto Esters

Asymmetric heterofunctionalization of $\beta$-keto esters has been a research interest of our group throughout the last few decades.[195, 209] More recently, mild hypervalent iodine reagents have been developed for the electrophilic trifluoromethylation of organic compounds.[210–214] Stereoselective trifluoromethylation has been principally achieved *via* the use of chiral auxiliaries;[215, 216] very recently, copper(II)-catalyzed trifluoromethylation of $\beta$-keto esters has been reported.[217]. Chiral titanium(IV)-[195] or ruthenium(II)-complexes[209] have shown to give high enantioselectivities in the asymmetric fluorination of $\beta$-keto esters. With a novel ruthenium(II) catalyst in our hands, its application in asymmetric trifluoromethylation was investigated (Table 3.15).

---

[‡]The borane protected sulfoxide **67** was kindly provided by Peter Ludwig.

Table 3.15: Attempted Ru$^{II}$-catalyzed asymmetric trifluoromethylation of $\beta$-keto esters.

| entry | R$^{1/2}$ | R$^3$ |
|---|---|---|
| 1 | Me / Me | 2,4,6-mesityl |
| 2 | $-(CH_2)_3-$ | Et |
| 3 | Ph / Me | Et |

In these preliminary reactions, the trifluoromethylation reagent[§] cleary interacted with the catalyst, as a color change from dark red to orange was observed shortly after the addition. While no product could be isolated, a cyclic ketone (entry 2, Table 3.15) showed a very small signal at $-70$ ppm, which could correspond to the $^{19}$F NMR shift of the desired product.[210] More thorough screening of reaction conditions was beyond the scope of this thesis.

---

[§]The reagent was kindly provided by Katrin Niedermann.

## 3.6 Conclusions

Monophosphino ferrocenes could be prepared *via* transmetallation of the prefiously described bromides **14–18** and **20** in 58–80% yield, giving overall yields of 36–64% starting from Ugi's amine (**1**).

The ligands could be coordinated to gold(I)- and ruthenium(II) precursors. Five complexes could be characterized by X-ray crystallography.

Gold(I) complex **51** showed no reactivity in intramolecular hydroamination of alkenes. Activation of the catalyst precursor *via* chloride abstraction could not be achieved. As an alternative, a methyl derivative was prepared which could be activate *via* protonation. However, it also showed no catalytic activity.

Ferrocenyl-tethered ruthenium(II) complexes **57**, **58** and **60** were applied in asymmetric transfer hydrogenation of acetophenone. The catalyses exhibited both base and counterion depencence. The best performance was achieved with all-phenyl catalyst **57**, giving 59% yield and an enantiomeric ratio of 72:28.

Ph−C(=O)−CH₃ →[5 mol-% **57**, 5 mol-% Et$_3$OSbF$_6$ / NH$i$-Pr$_2$, $i$-PrOH / 80 °C, 20 h] Ph−CH(OH)−CH₃, 59%, er 72:28

# Chapter 4

# General Conclusions and Outlook

Although the initial goal of this thesis, the synthesis of a ferrocenyl-tethered bifunctional ruthenium(II) catalyst for transfer hydrogenation was not achieved, the intermediate chiral ferrocenyl bromides were used as backbones for a variety of amino- and phosphino ferrocenes (Scheme 70). The monophosphino ferrocenes could be coordinated to gold(I)- and ruthenium(II)-precursors, giving the first ever ferrocenyl-tethered P–$\eta^6$-arene complex.

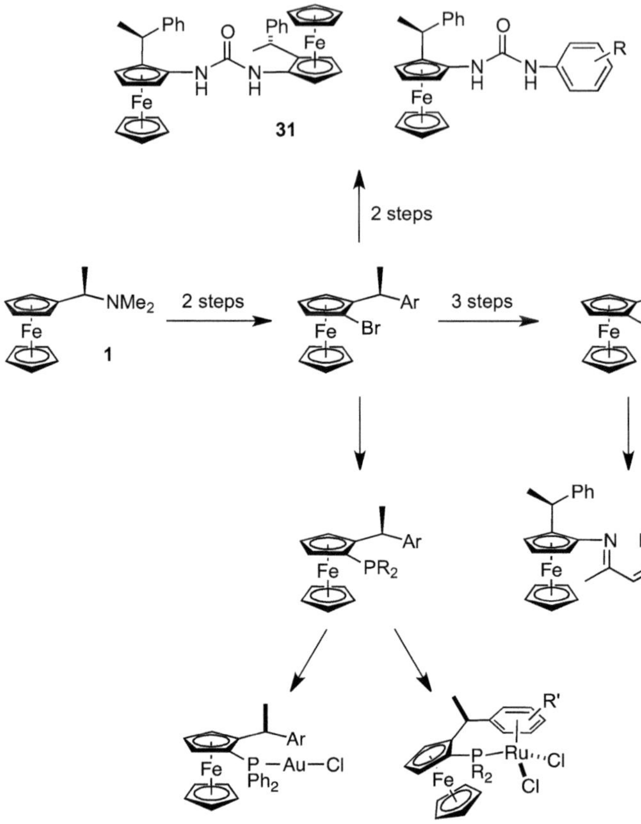

**Scheme 70:** Successfully synthesized amino- and phosphino ferrocenes, and their gold(I) and ruthenium(II) complexes.

## 4.1 Aminoferrocenes

The primary aminoferrocene **9** could be synthesized starting from Ugi's amine according to the intended route *via* Curtius rearrangement in good yields. As the attempted coordination of the amine to a series of metal precursors was unsuccessful, attempts were made to incorporate the aminoferrocene into a bidentate ligand. Functionalization of the amine proved to be very tedious, as its reactivity did not follow a clear pattern. Finally, Hartwig-Buchwald coupling and reductive amination reactions gave the best results.

Among other *N,N*-bidentate ligands, a chiral bisferrocenylNacNac ligand was prepared. At the time of their synthesis, bisferrocenylNacNac ligands were unknown to the literature and still, only one chiral NacNac ligand is known. Possibly due to the large steric demand of the ferrocenyl substituents, coordination as a bidentate ligand seems impossible.

Bisferrocenyl urea was also prepared. As a consequence, a range of monoferrocenyl ureas were synthesized and successfully applied as metallorganic organocatalysts. Preliminary results showed promising activities in the asymmetric allylation of hydrazones (Scheme 71).

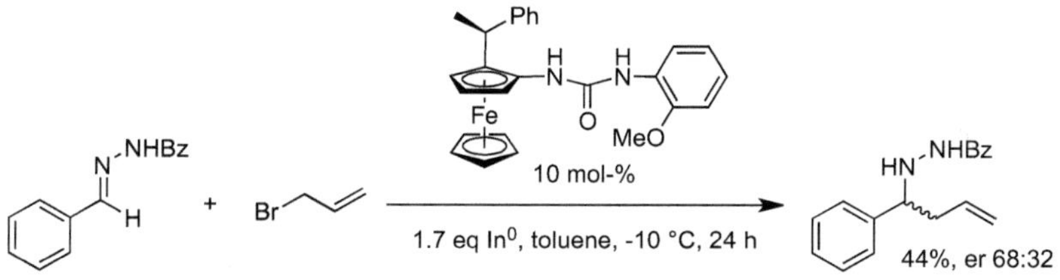

**Scheme 71:** Ferrocenyl urea-catalyzed allylation of hydrazones.

## 4.2 Phosphinoferrocenes

As the focus of this work on the synthesis of transition metal catalysts, the asymmetric ferrocenyl backbone was used for the synthesis of chiral phosphines. The corresponding gold(I) complexes could be prepared easily. Activating these compounds for their use in catalys *via* halide abstraction was not possible in a controlled fashion; potential solutions to these difficulties, like starting from cationic precursors or alkylation of the catalyst, faild to yield catalytically active complexes.

Finally, ferrocenyl-tethered ruthenium(II) complexes were prepared. With these compounds in hands, one of the initial objectives of this thesis was nearly reached—nearly, as no bifunctional ligands were prepared. The complexes were then applied in the transfer hydrogenation of acetophenone. The catalysts showed strong counterion effects. Also, their performance was also heavily influenced by the base added, but no clear pattern could be discerned. Overall, modest yields and stereoselectivities were obtained, the best result is shown in Scheme **4.3**. The low stereoselectivities were already suggested by the crystal structures. Although very different conformations of the ligands were observed, the ligands are oriented in a way that shows minimal steric influence to the coordination sphere of ruthenium. Consequently, the asymmetric induction during catalysis is slight.

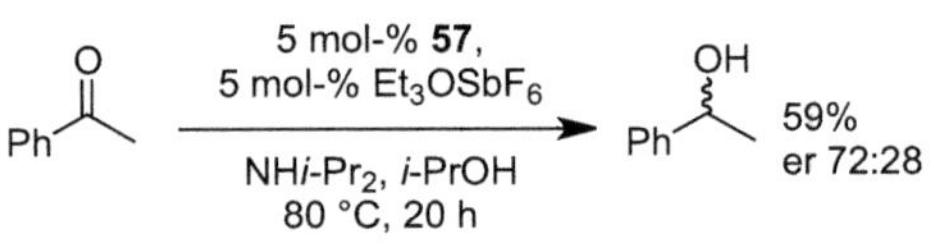

Ph–C(O)–CH$_3$ → [5 mol-% **57**, 5 mol-% Et$_3$OSbF$_6$, NH$i$-Pr$_2$, $i$-PrOH, 80 °C, 20 h] → Ph–CH(OH)–CH$_3$, 59%, er 72:28

**Scheme 72:** Our ferrocenyl-tethered catalysts gave only moderate results.

## 4.3 Outlook

Among the results obtained for the aminoferrocenes, the application of ferrocenyl ureas appear to be the most promising to investigate further. After all, its performance in asymmetric allylation was on par with Jacobsen's established ureas. Overall, our concept of an organometallic organocatalyst does not need to be reduced to monofunctionalized ferrocenes. Especially, varying the substituents on the sidechain, e. g. larger steric bulk or an additional hydrogen bond donor or -acceptor, could improve its stereoselectivity.

Even though the bisferrocenylNacNac ligand failed to yield transition metal complexes, its case does not need to be closed. While the bulk of two ferrocene moieties encumbered the projected coordination site, an asymmetric NacNac ligand bearing only one ferrocene may allow the formation of complexes while still exterting enough influence aver the active site to allow stereoselective applications.

Although chiral phosphinoferrocenes have been prepared before, our ligands are the first to contain both central and axial chirality, and a large field of application seems open for exploration. While successful applications of the gold(I) complexes prepared would appear to require a whole new paradigm, the ruthenium(II) complexes showed at least moderate activities in asymmetric transfer hydrogenation. According to the crystal structures, the metal fragment may be located both *endo* or *exo* with respect to the ferrocene. Thus, modifying the phosphine substituents would seem to be the most promising approach for future studies. Especially the preparation of *P*-chiral monophosphines might bring the source of chiralitiy much closer to the metal fragment.

# Chapter 5

# Experimental Part

## 5.1 General Remarks

Both (*R*)- and (*S*)-*N*,*N*-dimethyl-1-ferrocenylethylamine (**1**) were used as starting material. For convenience, reactions and analytics are only described for the derivatives of (*R*)-**1**.

### 5.1.1 Techniques

All manipulations with air or moisture sensitive materials were carried out at a vacuum/argon line using standard Schlenk techniques, or in a glovebox (MBRAUN MB-150B-GII or Lab Master 130) under an atmosphere of grade 5 nitrogen. Glassware were heated at 140 °C for a minimum of 2 h in an oven, dried under HV and subsequently put under argon.

### 5.1.2 Chemicals

(*R*)-Ugi's amine ((*R*)-**1**) was generously provided by Solvias AG (Basel) as tartrate salt. The following chemicals were prepared according to the literature: (*S*)-Ugi's amine ((*S*)-**1**),[27] [($\eta^6$-*p*-cymene)RuCl$_2$] (**68**),[218] diphenyl ditelluride,[113] [Ir($\eta^5$-Cp*)Cl$_2$]$_2$,[219] [Rh($\eta^5$-Cp*)Cl$_2$]$_2$,[219] KBAr$_F$,[220] HBAr$_F$,[221] Et$_3$OSbF$_6$,[222] (Me$_2$S)AuCl,[223]

[Cu(NCMe)$_4$]ClO$_4$,[224] Zr(NMe$_2$)$_4$,[225] [Ir(cod)Cl]$_2$,[226] 1-bromo-8-fluoronaphthalene,[227, 228] [Pd($\eta^3$-C$_3$H$_5$)Cl]$_2$,[229] 2,2-dimethylpent-4-enyl-1-amine,[230] benzyl 2,2-dimethylpent-4-enylcarbamate,[230] Ni(cod)$_2$,[231] (*E*)-*N*'-benzylidenebenzohydrazone,[232] *rac*-(*E*)-*N*'-(1-phenylbut-3-en-1-ylidene)benzohydrazide.[233] The following reagents were purified prior to use: acetylacetone (dist. from P$_2$O$_5$), chlorodiphenylphosphine (dist.), chlorodiiospropylphosphine (dist.), triethylamine (1 d dried over 4 Å molar sieves, then dist. from P$_2$O$_5$), allyl bromide (filtration through a plug of basic alox), potassium-*tert*-butoxide (sublimation). All solvents used for synthetic purposes were of puriss. p. a. grade, purchased from Fluka-Chemie AG or Armar Chemicals. Solvents used for extraction were of technical grade. The solvents for air- or moisture-sensitive manipulations were freshly distilled from an appropriate drying agent under argon (EtOH from Na/diethyl phthalate; toluene from Na; CH$_2$Cl$_2$ from CaH$_2$; Et$_2$O, THF, hexane and pentane from Na/benzophenone). DMF and dioxane were collected from a LC Technologiy Solutions Inc. SP-105 Solvent Purification System. Deuterated solvents for NMR spectroscopy were purchased from Cambridge Isotope Laboratories (CD$_2$Cl$_2$, toluene-*d*$_8$, CD$_3$CN, THF-*d*$_8$, C$_6$D$_6$, CD$_3$OD) or Armar Chemicals (CDCl$_3$). For sensitive compounds, CD$_2$Cl$_2$ was dried by heating under reflux for 12 h over CaH$_2$, then purified by distillation, degassed by three freeze-pump-thaw cycles and stored over 4 Å molar sieves; CDCl$_3$ was extracted with H$_2$O, dried over K$_2$CO$_3$, purified by distillation from P$_2$O$_5$, then degassed by three freeze-pump-thaw cycles and stored over 4 Å molar sieves. All other commercially available chemicals were obtained in puriss. p. a. grade from Fluka-Chemie AG, Aldrich-Fine Chemicals, Acros, ABCR-Chemicals, TCI-Deutschland GmbH or Strem Chemicals, and were used without further purification, unless stated otherwise.

## 5.1.3 Analytics

**Thin Layer Chromatography (TLC):** *Merck Silica Gel 60 $F_{254}$* visualized by fluorescence quenching at 254 nm. Additionally, TLC plates were stained using ninhydrine (0.6 g ninhydrine, 200 mL EtOH, 2 mL AcOH) or vanillin (12 g vanillin, 2 mL *conc.* $H_2SO_4$, 170 mL EtOH).

**Flash Chromatography:** Chromatographic purifications were performed on *Fluka Silica Gel 60* (230–400 mesh) using the given solvent ratios and a forced flow of eluent at 0.1–0.2 bar. Solvents were of technical grade and not distilled before use.

**NMR:** The $^1$H, $^{13}$C{1H}, $^{19}$F and $^{31}$P{$^1$H} NMR spectra were measured on the following instruments (frequencies in MHz): Bruker Avance AC 200 ($^1$H, 200.1; $^{19}$F, 188.3), Bruker Avance DPX 250 ($^1$H, 250.1; $^{13}$C{$^1$H}, 62.5; $^{31}$P{$^1$H}, 101.3), Bruker Avance DPX 300 ($^1$H, 300.1; $^{13}$C{$^1$H}, 75.5; $^{31}$P{$^1$H}, 121.5), Bruker Avance DPX 400 ($^1$H, 400.1; $^{13}$C{$^1$H}, 100.6; $^{31}$P{$^1$H}, 162.0), Bruker Avance DPX 500 ($^1$H, 500.2; $^{13}$C{$^1$H}, 125.8; $^{31}$P{$^1$H}, 202.5), and Bruker Avance DPX 700 ($^1$H, 700.1; $^{13}$C{$^1$H}, 176.0; $^{31}$P{1H}, 283.4). Two-dimensional NMR spectra were recorded on Bruker Avance DPX 300, DPX 500 or DPX 700 instruments. $^1$H and $^{13}$C chemical shifts ($\delta$) are expressed in parts per million (ppm) relative to tetramethylsilane as an external standard. $^{19}$F NMR signals are referenced to external $CFCl_3$, and $^{31}$P{$^1$H} NMR signals to external 85% $H_3PO_4$. Coupling constants $J$ are given in Hertz. The multiplicity is denoted by the following abbreviations: s: singlet; d: doublet; t: triplet; q: quartet; sept: septet; m: multiplet; *br*: broad. The following indices were used for assignment: ar: aromatic; bn: benzylic; Cp: substituted Cp-ring; Cp': unsubstituted Cp-ring. Complete assignment according to Figure 31.

**Chiral GC:** Enantiomeric ratios of acetophenone were determined by chiral GC on a Thermo Finnigan TraceGC ultra with a Thermo Finnigan AS2000 autosampler. Column: Supelco $\beta$-DEX 120 (30 m × 0.25 mm, film 0.25 $\mu$m). Inlet: Split injector (42 mL min$^{-1}$, 200 °C).

Carrier: Helium (1.4 mL min$^{-1}$). Detector: FID (Air/H$_2$ 350/35, 250 °C).

**HPLC**: Enantiomeric ratios of all other compounds were determined on a Hewlett-Packard 1050 Series or an Agilent 1100 Series with UV/VIS detection. The applied column (Daicel Chiralcel OD-H, AD-H or AM), flow rate (in mL min$^{-1}$), ratio of the eluents (*n*-hexane and 2-propanol), wavelength and sample injection volume (in μL, sample concentration 1 − 3 mg mL$^{-1}$) are specified for each compound.

**MS**: EI-, ESI- and MALDI-MS were measured by the MS-service of the Laboratory of Organic Chemistry (ETH Zurich).

**M.p.**: Melting points were determined on a Griffin MPA350 or a Büchi Melting Point B-540 apparatus in open capillaries and are uncorrected.

**EA**: Elemental analyses were carried out by the Laboratory of Microelemental Analysis (ETH Zurich).

**Optical rotation**: Measured at 589 nm (Na/Hal). Perkin Elmer 341 or Anton Parr MCP 200; cell 1 dm, solvent CHCl$_3$, c in g/100 mL.

**IR**: Infrared spectra were recorded on a Thermo Fisher Scientific Nicolet 6700 FT-IR equipped with a PIKE technologies GladiATR™ or a on a Perkin-Elmer BX II using ATR FT-IR technology and are reported as absorption maxima in cm$^{-1}$.

**Crystallography**: Intensity data of single crystals glued to a glass capillary were collected at the given temperature (usually 100 K) on a Bruker SMART APEX platform with CCD detector and graphite monochromated Mo-$K_\alpha$ radiation ($\lambda$ = 0.71073 Å). The program SMART served for data collection; integration was performed with the software SAINT [234]. The structures were solved by direct methods or Patterson methods, respectively, using the program SHELXS 97.[235] The refinement and all further calculations were carried out using SHELXL 97.[236] All non-hydrogen atoms were refined anisotropically using weighted full-matrix least squares on $F^2$. The hydrogen atoms were included in calculated positions and treated as riding atoms using SHELXL default parameters. In the end, absorption correction was applied (SADABS)[237] and weights were optimized in the final refinement cycles. The absolute configuration of chiral compounds was determined on the basis of the Flack parameter.[238, 239] The standard uncertainties (s.u.) are rounded according to the *Notes for Authors* of *Acta Crystallographica*.[240] Detailed crystallographic date are given in the Appendix.

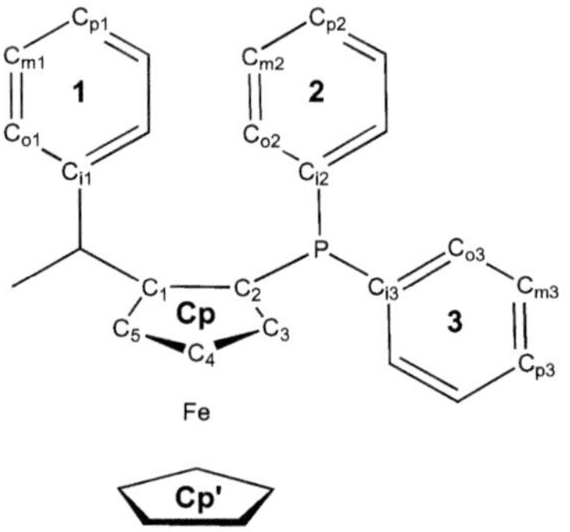

**Figure 31:** Atom indices in completely assigned molecules.

## 5.2 Syntheses

### 5.2.1 Ugi's Amine Resolution (1)

(R)-Ugi-amine tartrate (14.981 g, 36.8 mmol) was added to aq. NaOH (20%, 50 mL). The suspension was transferred to a separation funnel and extracted 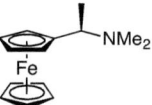 with Et$_2$O (3 × 50 mL). The combined organic phases were washed with aq. NaOH (20%, 50 mL), dried over Na$_2$SO$_4$, and the solvent was evaporated, leaving a red-brown oil which was stored at −18 °C. Yield: 9.289 g (98%), red-brown oil.
**$^1$H NMR** (250 MHz, CDCl$_3$): $\delta$ = 4.1–4.1 (m, 9H, CH$_{Cp}$ + CH$_{Cp'}$), 3.58 (q, $J$ = 6.9 Hz, 1H, C$H$(NMe$_2$)CH$_3$), 2.1 (s, 6H, NMe$_2$), 1.44 (d, $J$ = 6.9 Hz, 3H, CHC$H_3$).

### 5.2.2 ($S_p$)-1-Bromo-2-[($R$)-(1-$N,N$-dimethylamino)ethyl]ferrocene (12)

1 (2.248 g, 8.8 mmol) was dissolved in Et$_2$O (35 mL) and the solution was cooled to −78 °C. s-BuLi (1.3 M in cyclohexane:$n$-hexane 98:2; 8.75 mL, 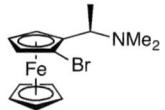 11.4 mmol, 1.3 eq) was added over 20 min keeping the temperature below −60 °C. The mixture was stirred for 30 min, the cooling bath removed and it was stirred for additional 60 min. The solution was then cooled again to −78 °C and a solution of 1,2-dibromotetrachloroethane (3.713 g, 11.4 mmol, 1.3 eq) in Et$_2$O (20 mL) was added slowly. The yellow suspension was allowed to warm to room temperature over night without the bath being removed, giving a pale brown suspension. NaHCO$_3$ (0.75 M, 45 mL) was added, giving a brown-red organic phase which was separated and an aqueous phase which was extracted with Et$_2$O (3 × 50 mL). The combined organic phases were washed with water (3 × 75 mL), dried over Na$_2$SO$_4$, and concentrated, leaving a red-brown oil (5.352 g) which was purified by flash chromatog-

raphy (EtOAc:*n*-hexane:NEt₃ 2:1:0.15). Yield: 2.740 g (93%), red-brown crystals.

**¹H NMR** (300 MHz, CDCl₃): $\delta$ = 4.46 (s, 1H, CH$_{Cp}$), 4.15 (s, 5H, CH$_{Cp'}$), 4.13 (s, 1H, CH$_{Cp}$), 4.10 (s, 1H, CH$_{Cp}$) 3.75 (q, $J$ = 6.9 Hz, 1H, C*H*(NMe₂)CH₃), 2.14 (s, 6H, NMe₂), 1.52 (d, $J$ = 6.9 Hz, 3H, CHC*H*₃).

### 5.2.3 ($S_p$)-1-Iodo-2-[($R$)-(1-*N*,*N*-dimethylamino)ethyl]ferrocene (13)

**1** (1.500 g, 5.8 mmol) was dissolved in Et₂O (15 mL) and the solution was cooled to −78 °C. *s*-BuLi (1.3 M in cyclohexane:*n*-hexane 98:2; 5.8 mL, 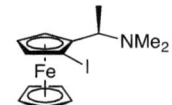 7.5 mmol, 1.29 eq) was added over 20 min, keeping the temperature below −60 °C. The mixture was stirred for 30 min, the cooling bath removed and stirred for additional 45 min. The solution was then cooled again to −78 °C and a solution of 1,2-diiodoethane (2.080 g, 7.4 mmol, 1.27 eq) in THF (3 mL) was added slowly. The yellow suspension was stirred at −78 °C for 15 min, then allowed to warm to 0 °C, and stirred for additional 20 min, giving a pale brown suspension. Aq. Na₂S₂O₃ (10 mL) was added, giving a brown-red organic phase which was separated and an aqueous phase which was extracted with Et₂O (3 × 10 mL). The combined organic phases were washed with water (3 × 25 mL), dried over Na₂SO₄, filtered and concentrated, leaving a red-brown oil (1.990 g). The residue was purified by flash chromatography (EtOAc:*n*-hexane:NEt₃ 2:1:0.15). Yield: 1.706 g (76%), red-brown crystals.

**¹H NMR** (300 MHz, CDCl₃): $\delta$ = 4.46 (dd, $J$ = 2.3, 1.3 Hz, 1H, CH$_{Cp}$), 4.24 (t, $J$ = 2.5 Hz, 1H, CH$_{Cp}$), 4.15 (m, 1H, CH$_{Cp}$), 4.12 (s, 5H, CH$_{Cp'}$), 3.62 (q, $J$ = 6.8 Hz, 1H, C*H*(NMe₂)CH₃), 2.14 (s, 6H, NMe₂), 1.50 (d, $J$ = 6.8 Hz, 3H, CHC*H*₃).

## 5.2.4 General Procedure for the Preparation of (1-Arylethyl)ferrocenes

*Best results were obtained starting from ≥1 mmol **12** and using freshly prepared Grignard-reagents.*
At −15 °C (NaCl/ice–bath) a solution of $ZnBr_2$ (~0.2 M in THF, 1.8 eq) was added dropwise to a solution of the corresponding aryl-Grignard-reagent (~0.2 M in THF, 1.6 eq). A white precipitate formed and the suspension was stirred for 30 min. The mixture was then cooled to −78 °C and a solution of **12** (~0.2 M in THF) was added dropwise, followed by addition of acetyl chloride (1.3 eq). The resulting yellow mixture was allowed to warm to room temperature overnight, giving a yellow suspension. $Et_2O$ was added and the reaction was quenched with water. The aqueous phase was extracted with $Et_2O$ (3 times) and brine (3 times) and dried over $MgSO_4$. Evaporation of the solvents gave the crude product which was purified by flash chromatography (n-pentane:$CH_2Cl_2$ 20:1) or recrystallization from EtOH.

## 5.2.5 ($S_p$)-1-Bromo-2-[(S)-1-phenylethyl]ferrocene (14)

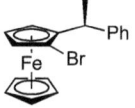

The reaction was carried out according to the general procedure (*vide supra*) using **12** (5.0 g), $ZnBr_2$ (6.0 g) and acetyl chloride (1.3 mL). The Grignard reagent was prepared from bromobenzene (2.45 mL) and magnesium turnings (861 mg) in THF (30 mL). Yield: 4.5 g (81%), orange crystals.
*When a commercially available solution of phenylmagnesiumbromide (Fluka, 1.0 M in THF) was used, 79% yield were reached.*
$^1$**H NMR** (400 MHz, $CDCl_3$): $\delta$ = 7.26 (m, 2H, $CH_{ar}$), 7.2–7.1 (m, 3H, $CH_{ar}$), 4.46 (dd, $J$ = 2.4, 1.4 Hz, 1H, $CH_{Cp}$), 4.31 (dd, $J$ = 2.3, 1.4 Hz, 1H, $CH_{Cp}$), 4.26 (s, 5H, $CH_{Cp'}$), 4.16 (t, $J$ = 2.5 Hz, 1H, $CH_{Cp}$), 4.10 (q, $J$ = 7.2 Hz, 1H, $CH(Ph)CH_3$), 1.72 (d, $J$ = 7.2 Hz, 3H, $CH_3$).
$^{13}$**C{$^1$H} NMR** (101 MHz, $CDCl_3$): $\delta$ = 146.7 (s, 1C, $C_{q,ar}$), 128.6

(s, 2C, CH$_{ar}$), 127.5 (s, 2C, CH$_{ar}$), 126.3 (s, 1C, CH$_{ar}$), 92.5 (s, 1C, C$_{q,Cp}$), 80.0 (s, 1C, C$_{q,Cp}$), 71.6 (s, 5C, CH$_{Cp'}$), 70.5 (s, 1C, CH$_{Cp}$), 65.8 (s, 1C, CH$_{Cp}$), 65.1 (s, 1C, CH$_{Cp}$), 38.9 (s, 1C, CH), 23.1 (s, 1C, CH$_3$).
**MS**: (HiRes EI-MS): calc: [M$^+$]: 367.9858; found: 367.9861 (100%).
**EA**: calc: C 58.58%, H 4.64%, Br 21.65%, Fe 15.13%; found: C 58.42%, H 4.64%.
**IR** (ATR-FT IR): $\tilde{\nu}$ = 3097 (w), 3080 (w), 3059 (w), 3024 (w), 2963 (w), 2926 (w), 2866 (w), 2359 (w), 1872 (w), 1633 (w), 1599 (w), 1582 (w), 1488 (m), 1447 (m), 1408 (w), 1384 (m), 1368 (m), 1344 (w), 1288 (w), 1246 (w), 1165 (w), 1104 (m), 1090 (m), 1076 (m), 1044 (w), 1024 (m), 999 (s), 965 (w), 938 (m), 908 (m), 890 (w), 874 (w), 840 (m), 819 (s), 765 (s), 718 (s), 697 (s), 639 (m), 616 (w).
$[\alpha]_D^{20}$ = −73.86 (c = 1, CHCl$_3$).
**Mp** = 145 °C.

## 5.2.6 ($S_p$)-1-Iodo-2-[($S$)-1-phenylethyl]ferrocene (15)

The reaction was carried out according to the general procedure (*vide supra*) using **13** (352.7 mg), ZnBr$_2$ (370 mg) and acetyl chloride (0.08 mL). The Grignard 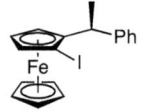 reagent was prepared from iodobenzene (370 mg) and $^i$PrMgCl (1.85 M in THF; 0.8 mL) in THF (5 mL). Yield: 237.1 mg (62%), orange solid.
**$^1$H NMR** (400 MHz, CDCl$_3$): $\delta$ = 7.26 (m, 2H, CH$_{ar}$), 7.2-7.1 (m, 3H, CH$_{ar}$), 4.47 (dd, $J$ = 2.3 Hz, $J$ = 1.3 Hz, 1H, CH$_{Cp}$), 4.38 (dd, $J$ = 2.2 Hz, $J$ = 1.2 Hz, 1H, CH$_{Cp}$), 4.28 (t, $J$ = 2.4 Hz, 1H, CH$_{Cp}$), 4.22 (s, 5H, CH$_{Cp'}$), 4.00 (q, $J$ = 7.2 Hz, 1H, C$H$(Ph)CH$_3$), 1.72 (d, $J$ = 7.2 Hz, 3H, CH$_3$).
**$^{13}$C{$^1$H} NMR** (101 MHz, CDCl$_3$): $\delta$ = 146.8 (s, 1C, C$_{q,ar}$), 128.6 (s, 2C, CH$_{ar}$), 127.7 (s, 2C, CH$_{ar}$), 126.3 (s, 1C, CH$_{ar}$), 94.6 (s, 1C, C$_{q,Cp}$), 74.9 (s, 1C, CH$_{Cp}$), 72.1 (s, 5C, CH$_{Cp'}$), 68.3 (s, 1C, CH$_{Cp}$), 65.8 (s, 1C, CH$_{Cp}$), 45.5 (s, 1C, C$_{q,Cp}$), 40.5 (s, 1C, CH), 23.4 (s, 1C,

$CH_3$).
**MS** (HiRes EI-MS): calc: [M+]: 415.9719; found: 415.9722 (100%).
**EA**: calc. C 51.96%, H 4.12%, Fe 13.42%, I 30.50%; found C 51.76%, H 4.07%.
**IR** (ATR-FT IR): $\tilde{\nu}$ = 3096 (w), 3022 (w), 2960 (m), 2925 (m), 2864 (w), 1873 (w), 1667 (w), 1598 (m), 1581 (w), 1488 (m), 1446 (m), 1409 (m), 1380 (m), 1366 (m), 1342 (m), 1282 (m), 1257 (w), 1240 (w), 1178 (w), 1106 (m), 1085 (m), 1076 (m), 1044 (w), 1026 (m), 999 (s), 930 (m), 908 (m), 891 (w), 872 (w), 840 (m), 816 (s), 778 (m), 760 (s), 720 (s), 697 (s), 665 (m), 638 (m), 614 (m).
$[\alpha]_D^{20}$ = −57.16 (c = 1, $CHCl_3$).
**Mp** = 100 °C.

## 5.2.7 ($S_p$)-1-Bromo-2-[($S$)-1-(3,5-dimethylphenyl)ethyl]ferrocene (16)

The reaction was carried out according to the general procedure (*vide supra*) using **12** (1.0 g), $ZnBr_2$ (1.2 g) and acetyl chloride (0.25 mL). The 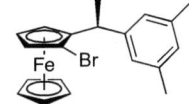 Grignard reagent was prepared from *m*-xylene-1-bromide (0.7 mL) and magnesium turnings (175 mg) in THF (10 mL). Yield: 1.2 g (99%), yellow solid.

**¹H NMR** (500 MHz, $CD_2Cl_2$): $\delta$ = 6.82 (s, 2H, $CH_{o1}$), 6.78 (d, 1H, $CH_{p1}$), 4.47 (dd, $J$ = 1.4, 1.0 Hz, 1H, $CH_{Cp,5}$), 4.36 (s, 1H, $CH_{Cp,3}$), 4.27 (s, 5H, $CH_{Cp'}$), 4.2 (t, 1H, $CH_{Cp,4}$), 4.02 (q, $J$ = 7.2 Hz, 1H, $CHCH_3$), 2.28 (s, 6H, $CH_{3,xyl}$), 1.67 (d, $J$ = 7.2 Hz, 3H, $CHCH_3$).
**¹³C{¹H} NMR** (126 MHz, $CD_2Cl_2$): $\delta$ = 146.8 (s, 1H, $C_{o1}$), 137.0 (s, 2C, $C_{m1}$), 127.9 (s, 1C, $C_{p1}$), 125.3 (s, 1C, $C_{o1}$), 92.6 (s, 1C, $C_{Cp,1}$), 79.8 (s, 1C, $C_{Cp,2}$), 71.6 (s, 5C, $C_{Cp'}$), 65.8 (s, 1C, $C_{Cp,4}$), 65.1 (s, 1C, $C_{Cp,3}$), 38.7 (s, 1C, $CHCH_3$), 23.0 (s, 1C, $CHCH_3$), 21.5 (s, 1C, $CH_{3,xyl}$).
**MS** (HiRes EI-MS): calc: [M+]: 396.0176; found: 396.0176 (100%).
**EA**: calc. C 60.49%, H 5.33%, Br 20.12%, Fe 14.06%; found C 60.35%,

H 5.34%, Br 20.04%.
**IR** (ATR-FT IR): $\tilde{\nu}$ = 2966 (m), 2913 (w), 1598 (m), 1448 (m), 1366 (m), 1240 (w).
$[\alpha]_D^{20}$ = −72.64 (c = 1, CHCl$_3$).
**Mp** = 75 °C.

### 5.2.8 ($S_p$)-1-Bromo-2-[($S$)-1-(2,4,6-trimethylphenyl)ethyl]ferrocene (17)

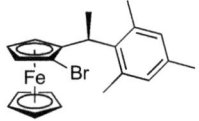

The reaction was carried out according to the general procedure (*vide supra*) using **12** (1.0 g), ZnBr$_2$ (1.2 g) and acetyl chloride (0.25 mL). The Grignard reagent was prepared from mesitylbromide (0.7 g) and magnesium turnings (175 mg) in THF (8 mL). Yield: 832 mg (68%), orange crystals.

**$^1$H NMR** (300 MHz, CD$_2$Cl$_2$): $\delta$ = 6.51 (s, 1H, CH$_{ar}$), 5.90 (s, 1H, CH$_{ar}$), 4.85 (q, $J$ = 7.3 Hz, 1H), 4.71 (s, 1H, CH$_{Cp}$), 4.27 (s, 1H, CH$_{Cp}$), 4.05 (s, 5H, CH$_{Cp'}$), 3.86 (s, 1H, CH$_{Cp}$), 2.60 (s, 3H, CH$_{3,mes}$), 1.91 (s, 3H, CH$_{3,mes}$), 1.72 (s, 3H, CH$_{3,mes}$), 1.61 (d, $J$ = 7.3 Hz, 3H, CH$_3$).

**$^{13}$C{$^1$H} NMR** (75 MHz, CD$_2$Cl$_2$): $\delta$ = 139.4 (s, 1C, C$_{ar}$), 135.8 (s, 1C, C$_{ar}$), 135.4 (s, 1C, C$_{ar}$), 135.2 (s, 1C, C$_{ar}$), 135.0 (s, 1C, C$_{ar}$), 130.9 (s, 1C, C$_{ar}$), 99.7 (s, 1C, C$_{Cp}$), 76.2 (s, 1C, C$_{Cp}$), 72.1 (s, 2C, C$_{Cp}$), 70.2 (s, 5C, C$_{Cp}$), 67.8 (s, 1C, C$_{Cp}$), 34.9 (s, 1C, CH), 21.6 (s, 1C, CH$_{3,mes}$), 21.3 (s, 1C, CH$_{3,mes}$), 20.5 (s, 1C, CH$_{3,mes}$), 17.4 (s, 1C, CH$_3$).

**MS** (HiRes EI-MS): calc: [M$^+$]: 410.0333; found: 410.0338 (100%).
**EA**: calc. C 61.35%, H 5.64%, Br 19.43%, Fe 13.58%; found C 61.29%, H 5.61%.
**Mp** = 98 °C.

## 5.2.9 ($S_p$)-1-Bromo-2-[($S$)-1-(3,5-di-*tert*-butylphenyl)ethyl]ferrocene (18)

The reaction was carried out according to the general procedure (*vide supra*) using **12** (300 mg), ZnBr$_2$ (360 mg) and acetyl chloride (0.08 mL). The Grignard reagent was prepared from 3,5-di-*tert*-butylbromobenzene (381 mg) and magnesium turnings (50 mg) in THF (5 mL). Yield: 306 mg (71%), orange oil.

**$^1$H NMR** (300 MHz, CD$_2$Cl$_2$): $\delta$ = 7.10–6.90 (m, 3H, CH$_{ar}$), 4.69 (s, 1H, CH$_{Cp}$), 4.43 (q, $J$ = 7.0 Hz, 1H, CH), 4.31 (s, 1H, CH$_{Cp}$), 4.13 (s, 5H, CH$_{Cp'}$), 3.76 (s, 1H, CH$_{Cp}$), 1.53 (d, $J$ = 7.0 Hz, 3H, CH$_3$), 1.02 (s, 18H, CH$_{3,t-Bu}$).

**$^{13}$C{$^1$H} NMR** (75 MHz, CD$_2$Cl$_2$): $\delta$ = 150.2 (s, 2C, C$_{ar}$), 146.5 (s, 1C, C$_{ar}$), 121.9 (s, 2C, C$_{ar}$), 119.6 (s, 1C, C$_{ar}$), 99.8 (s, 1C, C$_{Cp}$), 73.8 (d, 1C, C$_{Cp}$), 71.0 (s, 1C, C$_{Cp}$), 70.5 (s, 5C, C$_{Cp'}$), 70.0 (s, 1C, C$_{Cp}$), 69.5 (s, 1C, C$_{Cp}$), 38.4 (s, 1C, CH), 34.8 (s, 2C, C$_{t-Bu}$), 31.5 (s, 6C, CH$_{3,t-Bu}$), 24.5 (s, 1C, CH$_3$).

**MS** (HiRes EI-MS): calc: [M$^+$]: 480.1115; found: 480.1119 (100%).

**EA**: calc. C 64.88%, H 6.91%, Br 16.60%, Fe 11.60%; found C 64.86%, H 6.93%.

## 5.2.10 ($S_p$)-1-Bromo-2-[($S$)-1-(3,5-bis(trifluoromethyl)phenyl)ethyl]ferrocene (19)

The reaction was carried out according to the general procedure (*vide supra*) using **12** (1.65 g), ZnBr$_2$ (2 g) and acetyl chloride (0.4 mL). The Grignard reagent was prepared from 3,5-bis(trifluoromethyl)bromobenzene (1.46 mL) and magnesium turnings (289 mg) in THF (12 mL). Yield: 890 mg (36%), yellow crystals.

**¹H NMR** (500 MHz, CD$_2$Cl$_2$): $\delta$ = 7.73 (s, 1H, CH$_{p1}$), 7.64 (s, 2H, CH$_{m1}$), 4.53-4.26 (m, 3H, CH$_{Cp}$), 4.3 (s, 5H, CH$_{Cp'}$), 4.2 (s, 1H, CH$_{Cp,4}$), 4.28 (m, 1H, CHCH$_3$), 1.76 (d, $J_{H,H}$ = 7.3 Hz, 3H, CHCH$_3$).
**¹³C{¹H} NMR** (126 MHz, CD$_2$Cl$_2$): $\delta$ = 149.5 (s, 1C, C$_{i1}$), 127.9 (s, 2C, C$_{o1}$), 123.9 (q, $J$ = 271 Hz, 2C, CC$_3$), 120.5 (m, 2C, C$_{p1}$), 90.6 (s, 1C, C$_{Cp,2}$), 79.7 (s, 1C, C$_{Cp,1}$), 71.8 (s, 5C, C$_{Cp'}$), 70.9 (s, 1C, C$_{Cp,5}$), 66.4 (s, 1C, C$_{Cp,4}$), 65.0 (s, 1C, C$_{Cp,3}$), 39.0 (s, 1C, CHCH$_3$), 22.9 (s, 1C, CH$_3$).
**MS** (HiRes EI-MS): calc: [M$^+$]: 503.9611; found: 503.9615 (100%).
**EA**: calc. C 47.56%, H 2.99%, Br 15.82%, F 22.57%, Fe 11.06%; found C 47.64%, H 3.09%, Br 15.84%, F 22.45%.
**IR** (ATR-FT IR): $\tilde{\nu}$ = 3087 (w), 2974 (m), 2934 (w), 1619 (m), 1460 (m), 1369 (w), 1326 (m), 1168 (m).
$[\alpha]_D^{20}$ = $-77.41$ (c = 1, CHCl$_3$).
**Mp** = 91 °C.

## 5.2.11 ($S_p$)-1-Bromo-2-[($S$)-1-(1-naphthyl)ethyl]ferrocene (20)

The reaction was carried out according to the general procedure (*vide supra*) using **12** (2.0 g), ZnBr$_2$ (2.4 g) and acetyl chloride (0.5 mL). The Grignard reagent was prepared from 1-bromonaphthalene (1.3 g) and magnesium turnings (350 mg) in THF (15 mL). Yield: 1.8 g (72%), orange crystals.

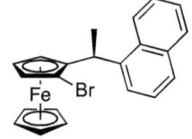

**¹H NMR** (300 MHz, CD$_2$Cl$_2$): $\delta$ = 8.24 (d, $J$ = 8.5 Hz, 1H, CH$_{ar}$), 7.77 (d, $J$ = 8.1 Hz, 1H, CH$_{ar}$), 7.56 (d, $J$ = 8.2 Hz, 1H, CH$_{ar}$), 7.50 (t, $J$ = 7.6 Hz, 1H, CH$_{ar}$), 7.40 (t, $J$ = 7.4 Hz, 1H, CH$_{ar}$), 7.18 (t, $J$ = 7.7 Hz, 1H, CH$_{ar}$), 6.82 (d, $J$ = 7.3 Hz, 1H, CH$_{ar}$), 4.84 (q, $J$ = 7.1 Hz, 1H, CH(Np)CH$_3$), 4.39 (s, 1H, CH$_{Cp}$), 4.36 (s, 1H, CH$_{Cp}$), 4.19 (s, 5H, CH$_{Cp'}$), 4.15 (s, 1H, CH$_{Cp}$), 1.66 (d, $J$ = 7.1 Hz, 3H, CH$_3$).
**¹³C{¹H} NMR** (75 MHz, CDCl$_3$): $\delta$ = 143.03 (s, 1C, C$_{q,ar}$), 133.78 (s, 1C, C$_{q,ar}$), 131.03 (s, 1C, C$_{q,ar}$), 128.98 (s, 1C, CH$_{ar}$), 126.52 (s,

1C, CH$_{ar}$), 125.87 (s, 1C, CH$_{ar}$), 125.60 (s, 1C, CH$_{ar}$), 125.35 (s, 1C, CH$_{ar}$), 123.82 (s, 1C, CH$_{ar}$), 123.23 (s, 1C, CH$_{ar}$), 92.33 (s, 1C, C$_{q,Cp}$), 79.83 (s, 1C, C$_{q,Cp}$), 71.31 (s, 5C, CH$_{Cp'}$), 70.25 (s, 1C, CH$_{Cp}$), 65.92 (s, 1C, CH$_{Cp}$), 65.37 (s, 1C, CH$_{Cp}$), 34.23 (s, 1C, CH), 21.96 (s, 1C, CH$_3$).
**MS** (HiRes EI-MS): calc: [M$^+$]: 418.0020; found: 418.0018 (100%).
**EA**: calc. C 63.04%, H 4.57%, Br 19.06%, Fe 13.32%; found C 63.08%, H 4.59%.
**Mp** = 125 °C.

## 5.2.12 (S$_p$)-1-Bromo-2-[(S)-1-(8-fluoronaphth-1-yl)ethyl]ferrocene (21)

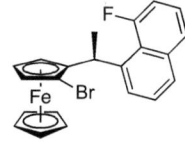

The reaction was carried out according to the general procedure (*vide supra*) using **12** (300 mg), ZnBr$_2$ (360 g) and acetyl chloride (0.08 mL). The Grignard reagent was prepared from 1-bromo-8-fluoronaphthalene (310 mg) and magnesium turnings (50 mg) in THF (5 mL). Yield: 184 mg (47%), orange crystals.
**$^1$H NMR** (300 MHz, CD$_2$Cl$_2$): $\delta$ = 7.57 (d, $J$ = 8.2 Hz, 2H, CH$_{ar}$), 7.32 (dd, $J$ = 12.8, 7.8 Hz, 1H, CH$_{ar}$), 7.22 – 7.09 (m, 2H, CH$_{ar}$), 6.84 (d, $J$ = 7.3 Hz, 1H, CH$_{ar}$), 5.19 (q, $J$ = 7.0 Hz, 1H, C$H$(Np)CH$_3$), 4.39 (s, 1H, CH$_{Cp}$), 4.34 (s, 1H, CH$_{Cp}$), 4.17 (s, 5H, CH$_{Cp'}$), 4.15 (s, 1H, CH$_{Cp}$), 1.64 (d, $J$ = 7.0 Hz, 3H, CH$_3$).
**$^{13}$C{$^1$H} NMR** (75 MHz, CD$_2$Cl$_2$): $\delta$ = 160.34 (d, $J$ = 252.3 Hz, 1C, CF), 143.02 (d, $J$ = 6.8 Hz, 1C, C$_{q,ar}$), 136.72 (d, $J$ = 5.3 Hz, 1C, C$_{q,ar}$), 126.68 (s, 1C, CH$_{ar}$), 126.65 (s, 1C, CH$_{ar}$), 125.92 (s, 1C, CH$_{ar}$), 125.64 (s, 1C, CH$_{ar}$), 125.53 (s, 1C, CH$_{ar}$), 125.48 (s, 1C, CH$_{ar}$), 111.87 (d, $J$ = 24.9 Hz, 1C, CH$_{ar}$), 92.87 (s, 1C, C$_{q,Cp}$), 80.45 (s, 1C, C$_{q,Cp}$), 71.71 (s, 5C, CH$_{Cp'}$), 70.59 (s, 1C, CH$_{Cp}$), 66.37 (s, 1C, CH$_{Cp}$), 65.72 (s, 1C, CH$_{Cp}$), 36.85 (d, $J$ = 15.2 Hz, 1C, CH$_3$), 22.77 (d, $J$ = 2.5 Hz, 1C, CH).
**$^{19}$F NMR** (282 MHz, CD$_2$Cl$_2$): $\delta$ = −112.94 (s).

**MS** (HiRes EI-MS): calc: [M$^+$]: 435.9925; found: 435.9927(100%).
**EA**: calc. C 60.45%, H 4.15%, Br 18.28%, F 4.35%, Fe 12.78%; found C 60.48%, H 4.14%, F 4.34%.
**Mp** = 121 °C.

## 5.2.13 ($S_p$)-Carboxyl-2-[($S$)-1-phenylethyl]ferrocene (7)

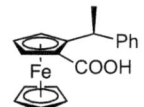

At −78 °C, $n$-BuLi (1.6 M in hexane; 5.1 mL, 8.2 mmol, 1.09 eq) was added to a solution of **14** (2.760 g, 7.5 mmol) in Et$_2$O (50 mL). After stirring for 30 min, the solution was allowed to warm to room temperature and stirred for additional 45 min. The red solution obtained was cooled to −78 °C and a stream of CO$_2$ was established. The color changed from bright red to pale orange. The mixture was stirred for 30 min at low temperature and another 30 min at room temperature, giving a pale brown suspension. The stream of CO$_2$ was then stopped and, in order to ensure completion of the reaction, a large excess of CO$_2$ (s) was added. The mixture was allowed to warm to room temperature. Aq. HCl (1 N, 50 mL) was added, the phases were separated, and the aqueous phase extracted with Et$_2$O (3 × 20 mL). The combined organic phases were washed with aq. HCl (1 N, 3 × 50 mL), dried over Na$_2$SO$_4$, filtered and the solvents were evaporated. The red solid was recrystallized from Et$_2$O/$n$-hexane. Yield: 2.440 g (98%), red crystals.
*This compound was also prepared in a similar fashion starting from* **15**.
$^1$**H NMR** (300 MHz, CDCl$_3$): $\delta$ = 12.1–10.5 (b, 1H, COO$H$), 7.2–7.1 (m, 5H, C$H_{ar}$), 4.83 (dd, $J$ = 2.5, 1.5 Hz, 1H, C$H_{Cp}$), 4.7 4.6 (m, 2H, C$H_{Cp}$ + C$H$(Ph)CH$_3$), 4.46 (t, $J$ = 2.6 Hz, 1H, C$H_{Cp}$), 4.26 (s, 5H, C$H_{Cp'}$), 1.69 (d, $J$ = 7.3 Hz, 3H, CH$_3$).
$^{13}$**C{$^1$H} NMR** (75 MHz, CDCl$_3$): $\delta$ = 177.9 (s, 1C, COOH), 147.6 (s, 1C, C$_{q,ar}$), 128.0 (s, 2C, CH$_{ar}$), 127.0 (s, 2C, CH$_{ar}$), 125.7 (s, 1C,

CH$_{ar}$), 97.3 (s, 1C, C$_{q,Cp}$), 71.3 (s, 1C, CH$_{Cp}$), 70.7 (m, 5C, CH$_{Cp'}$), 70.6 (s, 1C, CH$_{Cp}$), 69.9 (s, 1C, CH$_{Cp}$), 67.0 (s, 1C, C$_{q,Cp}$), 37.4 (s, 1C, CH), 22.9 (s, 1C, CH$_3$).
**MS** (HiRes EI-MS): calc: [M$^+$]: 334.0651; found: 334.0654 (100%).
**EA** calc. C 68.29%, H 5.43%, Fe 16.71%, O 9.57%; found C 68.30%, H 5.55%.
**IR** (ATR-FT IR) $\tilde{\nu}$ = 2962 (w), 2599 (w), 1665 (s), 1603 (w), 1492 (m), 1467 (s), 1412 (w), 1368 (m), 1352 (w), 1303 (s), 1272 (m), 1230 (s), 1194 (w), 1106 (w), 1098 (m), 1050 (w), 1027 (w), 1000 (s), 967 (m) 929 (m), 8878 (m), 856 (m), 824 (s), 813 (s), 781 (m), 763 (s), 708 (s), 697 (s), 650 (m), 618 (w).
$[\alpha]_D^{20} = -124.23$ (c = 1, CHCl$_3$).
**Mp** 132 °C.

## 5.2.14 ($S_p$)-$N$-(Benzyloxycarbonylamino)-2-[($S$)-1-phenylethyl]ferrocene (8)

The reaction was monitored by TLC (n-hexane/EtOAc 3:1). $R_f$ values and spot colors: azide 0.7 (bright red; brown with vanillin), isocyanate 0.9 (purple with vanillin), carbamate 0.6 (orange; brown with vanillin).
NEt$_3$ (1.5 mL, 10.8 mmol, 1.5 eq) was added to a dark-red solution of the carboxylic acid **7** (2.410 g, 7.2 mmol) in toluene (12 mL) and the mixture was stirred for 10 min, clearing up slightly. dppa (1.75 mL, 8. mmol, 1.13 eq) was added, leading to a darkening of the mixture and it was stirred at room temperature for 25 min, then heated to 90 °C. Upon heating, the evolution of N$_2$ was observed. After 30 min, benzylic alcohol (1.5 mL, 14.6 mmol, 2.02 eq) was added. The mixture was stirred for another 15 min at 90 °C, then it was cooled to room temperature. Water (30 mL) and Et$_2$O (20 mL) was added and the phases were separated. The aqueous phase was extracted with Et$_2$O (3 × 20 mL). The combined organic phases were washed with

water (3 × 30 mL), dried over MgSO$_4$ and the solvent was evaporated leaving a red oil which was purified by column chromatography (*n*-hexane:EtOAc 7:1).

Yield: 2.943 g (93%), orange-red solid.

**$^1$H NMR** (300 MHz, CDCl$_3$): $\delta =$ 7.4–7.1 (m, 10H, C$H_{ar}$), 5.47 (b, 1H, NH), 5.09 (b, 2H, CH$_{Cp}$), 4.78 (b, 1H, CH$_{Cp}$), 4.18 (s, 6H, CH$_{Cp'}$ + CH$_{bn}$), 4.1–4.0 (m, 2H, C$H$(Ph)CH$_3$ + CH$_{bn}$), 1.66 (d, $J =$ 7.1 Hz, 3H, CH$_3$).

**$^{13}$C{$^1$H} NMR** (75 MHz, CDCl$_3$): $\delta =$ 145.7 (s, 1C, C$_{Cbz}$), 136.3 (s, 1C, C$_{q,ar}$), 128.7 (s, 1C, CH$_{ar}$), 128.5 (s, 1C, CH$_{ar}$), 128.2 (s, 1C, CH$_{ar}$), 126.8 (s, 1C, CH$_{ar}$), 126.4 (s, 1C, CH$_{ar}$), 77.3 (s, 1C, C$_{q,Cp}$), 69.8 (s, 5C, CH$_{Cp'}$), 66.8 (b, 2C, C$_{Cp}$), 63.0 (s, 1C, C$_{bn}$), 61.5 (b, 1C, C$_{Cp}$), 38.2 (s, 1C, $C$H(Ph)CH$_3$), 22.7 (s, 1C, CH$_3$)

**MS** (HiRes EI-MS): calc. [M$^+$]: 439.1229; found: 439.1236 (25.4%).

**EA**: calc. C 71.08%, H 5.74%, Fe 12.71%, N 3.19%, O 7.28%; found: C 71.04%, H 5.83%, N 3.21%.

**IR** (ATR-FT IR): $\tilde{\nu} =$ 3204 (w), 3101 (w), 2968 (w), 2929 (w), 2362 (w), 1706 (s), 1600 (w), 1490 (m), 1443 (s), 1398 (s), 1339 (s), 1330 (s), 1286 (m), 1211 (w), 1160 (w), 1113 (w), 1104 (w), 1060 (s), 1050 (s), 1029 (s), 997 (m), 916 (w), 834 (w), 816 (m), 798 (m), 758 (w), 728 (s), 701 (s), 694 (s), 642 (m), 610 (m).

$[\alpha]_D^{20} = -496.39$ (c = 1, CHCl$_3$).

**Mp** = 82 °C.

## 5.2.15 ($S_p$)-2-[($S$)-1-phenylethyl]aminoferrocene (9)

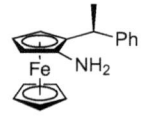

*i*-PrOH and H$_2$O were degassed prior to use. The carbamate **8** (2.650 g, 6.03 mmol) was dissolved in *i*-PrOH (45 mL). H$_2$O (30 mL) and KOH (16.7 g, 297 mmol) were added and the suspension was heated to reflux over night. The phases were separated, the aqueous phase was extracted with Et$_2$O (3 × 20 mL), the combined organic phases were washed with aq. NaOH (10%, 3 × 30 mL), dried over Na$_2$SO$_4$ and the solvent was evaporated, giving an orange-red solid which was purified by flash chromatography (n-hexane:EtOAc 3:1, 1.5% NEt$_3$). Traces of benzylic alcohol were removed under HV. Yield: 1.672 g (91%), orange microcrystals.

**$^1$H NMR** (300 MHz, CDCl$_3$): $\delta$ = 7.3–7.2 (m, 2H, CH$_{ar}$), 7.2–7.11 (m, 3H, CH$_{ar}$), 4.11 (s, 5H, CH$_{Cp'}$), 4.1–4.0 (m, 2H, C$H$(Ph)CH$_3$ + CH$_{Cp}$), 4.00 (s, 1H, CH$_{Cp}$), 3.85 (s, 1H, CH$_{Cp}$), 2.29 (s, 2H, NH$_2$), 1.67 (d, $J$ = 7.1 Hz, 3H, CH$_3$).

**$^{13}$C{$^1$H} NMR** (75 MHz, CDCl$_3$): $\delta$ = 146.4 (s, 1C, C$_{q,ar}$), 128.5 (s, 2C, CH$_{ar}$), 126.8 (s, 2C, CH$_{ar}$), 126.0 (s, 1C, CH$_{ar}$), 82.7 (s, 1C, C$_{q,Cp}$), 76.6 (s, 1C, C$_{q,Cp}$), 69.5 (s, 5C, CH$_{Cp'}$), 62.4 (s, 1C, CH$_{Cp}$), 60.9 (s, 1C, CH$_{Cp}$), 59.0 (s, 1C, CH$_{Cp}$), 38.2 (s, 1C, CH), 22.6 (s, 1C, CH$_3$).

**MS** (HiRes EI-MS): calc: [M$^+$]: 305.0861; found: 305.0860 (100%).

**EA**: calc. C 70.84%, H 6.27%, Fe 18.30%, N 4.59%, ; found C 70.65%, H 6.37%, N 4.52%

**IR** (ATR-FT IR): $\tilde{\nu}$ = 3431 (w), 3092 (w), 3022 (w), 2971 (m), 2929 (w), 2869 (w), 1875 (w), 1732 (w), 1600 (m), 1472 (m), 1446 (s), 1408 (m), 1367 (m), 1307 (w), 1258 (w), 1220 (w), 1187 (w), 1154 (w), 1101 (s), 1079 (w), 1050 (w), 1029 (m), 997 (s), 966 (w), 910 (w), 878 (w), 836 (m), 811 (s), 769 (m), 731 (m), 700 (s), 651 (m), 620 (m).

$[\alpha]_D^{20}$ = −100.28 (c = 1, CHCl$_3$).

**Mp** = 128 °C.

## 5.2.16 (E)-N-(((Z)-4-((S$_p$)-2-[(S)-1-phenylethyl]ferrocen-1-yl)amino)pent-3-en-2-yliden)-(S$_p$)-2-[(S)-1-phenylethyl]-aminoferrocene (29)

In a two-necked round bottomed flask equipped with a Dean-Stark-trap, a solution of aminoferrocene **9** (2.5 g, 8.2 mmol), acetylacetone (0.83 mL, 8.2 mmol, 1 eq) and p-TsOH · H$_2$O (1.6 g, 8.4 mmol, 1.05 eq) in toluene (250 mL) was stirred at reflux for 5 h. Complete consumtion of the amine was confirmed by TLC (n-hexane:EtOAc 3:1, 2.5% NEt$_3$, staining with ninhydrine). The solution was allowed to cool to r. t. and an additional equivalent of **9** was added. The mixture was then stirred at reflux for 40 h. After consumption of the amine had been confirmed by TLC, the solution was allowed to cool to r. t. and poured into aq. NaOH (0.5 M, 250 mL). The phases were separated, the organic phase was dried over Na$_2$SO$_4$ and the solvent was removed under reduced pressure. The crude product was purified by recrystallisation from acetone at −18 °C. Yield: 4.4 g (80%), red crystals.

**$^1$H NMR** (300 MHz, CD$_2$Cl$_2$): $\delta$ = 11.22 (s, 1H, NH), 7.18 (m, 4H, CH$_{ar}$), 7.14 (m, 4H, CH$_{ar}$), 7.01 (m, 2H, CH$_{ar}$), 4.42 (s, 10H, CH$_{Cp'}$), 4.37 (s, 1H, CH$_{Nac}$), 4.29 (s, 2H, CH$_{Cp}$), 4.24 (s, 4H, CH$_{Cp}$), 4.22 (q, J = 7.1 Hz, 2H, CH(Ph)CH$_3$), 4.10 (s, 2H, CH$_{Cp}$), 1.75 (d, J = 7.2 Hz, 6H, CHCH$_3$), 1.17 (s, 6H, CH$_{3,Nac}$).

**$^{13}$C{$^1$H} NMR** (75 MHz, CD$_2$Cl$_2$): $\delta$ = 163.17 (s, 2C, C$_{q,Nac}$), 147.54 (s, 2C, C$_{q,ar}$), 128.35 (s, 4C, CH$_{ar}$), 127.36 (s, 4C, CH$_{ar}$), 125.95 (s, 2C, CH$_{ar}$), 101.85 (s, 2C, C$_{q,Cp}$), 96.89 (s, 1C, CH$_{Nac}$), 89.29 (s, 2C, C$_{q,Cp}$), 70.08 (s, 10C, CH$_{Cp'}$), 65.63 (s, 2C, CH$_{Cp}$), 63.16 (s, 2C, CH$_{Cp}$), 62.98 (s, 2C, CH$_{Cp}$), 37.99 (s, 2C, CH(Ph)CH$_3$), 21.81 (s, 2C, CHCH$_3$), 20.57 (s, 2C, CH$_{3,Nac}$).

**MS** (HiRes EI-MS): calc: [M$^+$]: 674.2047; found: 674.2051 (100%).

**EA:** calc. C 73.01%, H 6.28%, Fe 16.56%, N 4.15%; found C 73.05%, H 6.35%, N 4.11%.
**Mp** = 186 °C.

### 5.2.17 N,N'-bis((S$_p$)-2-[(S)-1-phenylethyl]ferrocen-1-yl)urea (31)

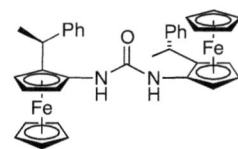

The reaction was carried out similar to the preparation of **8** using acid **7** (301.7 mg, 0.9 mmol), NEt$_3$ (0.30 mL, 2.16 mmol, 2.4 eq) and dppa (0.22 mL, 1.02 mmol, 1.13 eq). Instead of BnOH, the isocyanate was quenched with water (0.05 mL, 2.78 mmol, 3.07 eq). Yield: 291 mg (quant), orange solid.

**$^1$H NMR** (700 MHz, CDCl$_3$): $\delta$ = 6.99 (s, 6H, CH$_{ar}$), 6.77 (s, 4H, CH$_{ar}$), 5.64 (s, 2H, NH), 4.36 (s, 2H, CH$_{Cp}$), 4.25 (s, 10H, CH$_{Cp'}$), 4.14 (s, 2H, CH$_{Cp}$), 4.03 (s, 2H, CH$_{Cp}$), 3.83 (s, 2H, CH(Ph)CH$_3$), 1.56 (s, 6H, CH$_3$).
**$^{13}$C{$^1$H} NMR** (176 MHz, CDCl$_3$): $\delta$ = 155.0 (s, 1C, C$_{urea}$), 145.6 (s, 2C, C$_{q,ar}$), 128.4 (s, 4C, CH$_{ar}$), 126.7 (s, 4C, CH$_{ar}$), 126.0 (s, 2C, CH$_{ar}$), 92.5 (s, 2C, C$_{q,Cp}$), 87.1 (s, 2C, C$_{q,Cp}$), 69.9 (s, 10C, CH$_{Cp'}$), 63.5 (s, 6C, CH$_{Cp}$), 38.1 (s, 2C, CH), 22.5 (s, 2C, CH$_3$).
**MS** (HiRes EI-MS): calc: [M$^+$]: 637.1600; found: 637.1616 (100%).
**Mp** = 104 °C (decomp).

### 5.2.18 N-(S$_p$)-2-[(S)-1-phenylethyl]ferrocen-1-yl)-N'-phenylurea (32)

The reaction was carried out similar to the preparation of **8** using acid **7** (295.9 mg, 0.9 mmol), NEt$_3$ (0.29 mL, 2.09 mmol, 2.4 eq) and dppa (0.22 mL, 1.02 mmol, 1.15 eq). Instead of BnOH, the isocyanate was quenched with aniline (0.10 mL, 1.10 mmol, 1.24 eq). Yield: 261 mg (69%), orange solid.

**¹H NMR** (300 MHz, CDCl₃): δ = 7.14 – 6.99 (m, 6H, CH$_{ar}$), 6.89 (b, 4H, CH$_{ar}$), 6.19 (s, 1H, NH), 6.02 (s, 1H, NH), 4.36 (s, 1H, CH$_{Cp}$), 4.31 (s, 1H, CH$_{Cp}$), 4.21 (s, 5H, CH$_{Cp'}$), 4.08 (s, 1H, CH$_{Cp}$), 3.93 (dd, $J$ = 7.0 Hz, 1H, CH), 1.58 (d, $J$ = 7.9 Hz, 3H, CH₃).
**¹³C{¹H} NMR** (75 MHz, CDCl₃): δ = 154.8 (s, 1C, C$_{urea}$), 146.1 (s, 1C, C$_{q,Cp}$), 138.0 (s, 1C, C$_{q,ar}$), 128.8 (s, 2C, CH$_{ar}$), 128.7 (s, 2C, CH$_{ar}$), 126.7 (s, 2C, CH$_{ar}$), 126.4 (s, 1C, CH$_{ar}$), 123.2 (s, 1C, CH$_{ar}$), 119.9 (s, 2C, CH$_{ar}$), 70.2 (s, 5C, CH$_{Cp'}$), 64.6 (s, 3H, CH$_{Cp}$), 37.9 (s, 2H, CH), 22.6 (s, 1H, CH₃).
**MS** (HiRes EI-MS): calc: [M⁺]: 424.1233; found: 424.1223 (100%).
**EA**: calc. C 70.77%, H 5.70%, Fe 13.16%, N 6.60%, O 3.77%; found C 70.56%, H 5.81%, N 6.58%.
**[α]$_D^{20}$** = −290 (c = 1, CHCl₃).
**Mp** = 193 °C (decomp).

### 5.2.19 *N*-(*S$_p$*)-2-[(*S*)-1-phenylethyl]ferrocen-1-yl)-*N'*-(2-methoxyphenyl)urea (33)

The reaction was carried out similar to the preparation of **8** using acid **7** (100.1 mg, 0.3 mmol), NEt₃ (0.09 mL, 0.65 mmol, 2.17 eq) and dppa (0.22 mL, 1.02 mmol, 3.41 eq). Instead of BnOH, the isocyanate was quenched with *ortho*-anisidine (35 μL, 0.13 mmol, 1.04 eq). Yield: 99 mg (73%), orange solid.
**¹H NMR** (300 MHz, CDCl₃): δ = 7.86 (d, $J$ = 7.7 Hz, 1H, CH$_{ar}$), 7.07–6.96 (m, 5H, CH$_{ar}$), 6.91–6.74 (m, 3H, CH$_{ar}$), 6.67 (s, 1H, CH$_{ar}$), 5.59 (s, 1H, NH), 5.22 (s, 1H, NH), 4.43 (s, 1H, CH$_{Cp}$), 4.31 (s, 1H, CH$_{Cp}$), 4.22 (s, 5H, CH$_{Cp'}$), 4.11 (s, 1H, CH$_{Cp}$), 3.89 (s, 1H, C*H*CH₃), 3.63 (s, 3H, OCH₃), 1.55 (d, $J$ = 7.0 Hz, 3H, CH₃).
**¹³C{¹H} NMR** (75 MHz, CDCl₃): δ = 154.1 (s, 1C, C$_{urea}$), 148.1 (s, 1C, C$_{q,ar}$), 128.5 (s, 2C, CH$_{ar}$), 128.0 (s, 1C, C$_{q,ar}$), 126.7 (s, 2C, CH$_{ar}$), 126.2 (s, 1C, CH$_{ar}$), 122.5 (s, 1C, CH$_{ar}$), 121.0 (s, 1C, CH$_{ar}$), 119.5 (s, 1C, CH$_{ar}$), 109.8 (s, 1C, CH$_{ar}$), 70.3 (s, 5C, CH$_{Cp'}$), 64.2 (s,

3C, CH$_{Cp}$), 55.5 (s, 1C, OCH$_3$), 38.0 (s, 1C, $C$HCH$_3$), 22.4 (s, 1C, CH$C$H$_3$).
**MS** (HiRes EI-MS): calc: [M$^+$]: 454.1344; found: 454.1346 (100%).
**EA**: calc. C 68.73%, H 5.77%, Fe 12.29%, N 6.16%, O 7.04%; found C 68.44%, H 6.05%, N 5.90%.
$[\alpha]_D^{20} = -385$ (c = 1, CHCl$_3$).
**Mp** = 113–116 °C.

## 5.2.20  $N$-($S_p$)-2-[($S$)-1-phenylethyl]ferrocen-1-yl)-$N'$-(6-methylpyridin-2-yl)urea (34)

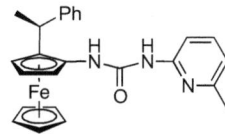

The reaction was carried out similar to the preparation of **8** using acid **7** (101.4 mg, 0.3 mmol), NEt$_3$ (0.09 mL, 65 mmol, 2.14 eq) and dppa (0.22 mL, 1.02 mmol, 3.36 eq). Instead of BnOH, the isocyanate was quenched with 2-amino-6-methylpyridine (37.1 mg, 0.34 mmol, 1.13 eq). Yield: 116 mg (84%), orange solid.
**$^1$H NMR** (300 MHz, CDCl$_3$): $\delta$ = 10.98 (s, 1H, NH), 8.60 (s, 1H, NH), 7.40 (t, $J$ = 7.8 Hz, 1H, CH$_{ar}$), 6.99 (s, 4H, CH$_{ar}$), 6.91 (s, 1H, CH$_{ar}$), 6.67 (d, $J$ = 7.3 Hz, 1H, CH$_{ar}$), 6.57 (s, 1H, CH$_{Cp}$), 4.91 (s, 1H, CH$_{Cp}$), 4.13 (s, 5H, CH$_{Cp'}$), 4.06 (s, 1H, C$H$CH$_3$), 4.03 (s, 1H, CH$_{Cp}$), 3.99 (s, 1H, CH$_{Cp}$), 2.42 (s, 3H, CH$_3$), 1.69 (d, $J$ = 7.0 Hz, 3H, CHC$H_3$).
**$^{13}$C{$^1$H} NMR** (75 MHz, CDCl$_3$): $\delta$ = 155.0 (s, 1C, C$_{urea}$), 154.4 (s, 1C, $C$CH$_{3,pic}$), 152.7 (s, 1C, C$_{q,pic}$), 146.0 (s, 1C, C$_{q,ar}$), 138.7 (s, 1C, CH$_{ar}$), 128.2 (s, 2C, CH$_{ar}$), 126.8 (s, 2C, CH$_{ar}$), 126.0 (s, 1C, CH$_{ar}$), 116.3 (s, 1C, CH$_{ar}$), 109.2 (s, 1C, CH$_{ar}$), 85.9 (b, 1C, C$_{q,Cp}$), 69.8 (s, 5C, CH$_{Cp'}$), 65.0 (s, 1C, CH$_{Cp}$), 63.6 (s, 1C, CH$_{Cp}$), 63.1 (s, 1C, CH$_{Cp}$), 37.3 (s, 1C, $C$HCH$_3$), 24.3 (s, 1C, CH$_{3,pic}$), 22.4 (s, 1C, CH$C$H$_3$).
**MS** (HiRes EI-MS): calc: [M$^+$]: 439.1342; found: 439.1339 (100%).
**EA**: calc. C 68.35%, H 5.74%, Fe 12.71%, N 9.56%, O 3.64%; found C 68.31%, H 5.85%, N 9.57%.
$[\alpha]_D^{20} = -103.47$ (c = 1, CHCl$_3$).

Mp = 170 °C.

## 5.2.21 ($S_p$)-1-Formyl-2-[($S$)-1-phenylethyl]ferrocene (39)

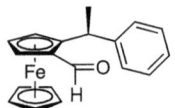

A solution of bromide **14** (260 mg, 0.7 mmol) in Et$_2$O (4 mL) was cooled to −78 °C and $n$-BuLi (1.6 M in hexane, 0.5 mL, 0.8 mmol, 1.15 eq) was added. After 20 min, the mixture was allowed to warm to r. t. After 30 min, the solution was cooled back to −78 °C, DMF (0.1 mL, 2.0 mmol, 2.8 eq) was added and the mixture was allowed to warm to r. t. over 10 min, then quenched with water (10 mL). The phases were separated, the aqueous phase was extracted with Et$_2$O and the combined organic phases were dried over Na$_2$SO$_4$. Evaporation of the solvents gave the pure product. Yield: 221 mg (quant), red crystals.

**$^1$H NMR** (300 MHz, CDCl$_3$): $\delta$ = 9.92 (s, 1H, CHO), 7.16 – 7.07 (m, 2H, CH$_{ar}$), 7.06 – 6.98 (m, 3H, CH$_{ar}$), 4.68 (s, 2H, CH$_{Cp}$), 4.52 (s, 1H, CH$_{Cp}$), 4.33 (q, $J$ = 7.2 Hz, 1H, C$H$(Ph)CH$_3$), 4.22 (s, 5H, CH$_{Cp'}$), 1.62 (d, $J$ = 7.2 Hz, 3H, CH$_3$).

**$^{13}$C{$^1$H} NMR** (75 MHz, CDCl$_3$): $\delta$ = 193.4 (s, 1C, CHO), 147.4 (s, 1C, C$_{q,ar}$), 128.4 (s, 1C, CH$_{ar}$), 126.8 (s, 2C, CH$_{ar}$), 126.2 (s, 1C, CH$_{ar}$), 97.0 (s, 1C, C$_{q,Cp}$CHO), 76.4 (s, 1C, C$_{q,Cp}$), 71.6 (s, 1H, CH$_{Cp}$), 71.2 (s, 1C, CH$_{Cp}$), 70.30 (s, 5C, CH$_{Cp'}$), 70.1 (s, 1C, CH$_{Cp}$), 38.0 (s, 1C, CH), 22.44 (s, 1C, CH$_3$)

**MS** (HiRes EI-MS): calc. [M$^+$]: 318.0707; found: 318.0715 (100%).

**EA**: calc. C 71.72%, H 5.70%, Fe 17.55%, O 5.03%; found C 71.82%, H 5.71%.

**Mp** = 136 °C.

## 5.2.22 ($S_p$)-1-Diphenylphosphino-2-[($S$)-1-phenylethyl]ferrocene (40)

A solution of **14** (2.00 g, 5.44 mmol) in THF (40 mL) was cooled to −78 °C. n-BuLi (1.6 M in hexane, 3.8 mL, 6.08 mmol, 1.1 eq) was added, the color of the solution turning from orange to red. After stirring for 30 min, the cooling bath was removed and stirring was continued for another hour. The solution was again cooled to −78 °C and chloro diphenylphosphine (3 mL, 16.7 mmol, 3.1 eq) was added dropwise, the solution turning from red to yellow. After stirring for 30 min, the cooling bath was removed and the reaction mixture was heated to reflux overnight, and the solution turned red again. The solution was then quenched with sat. $NaHCO_3$, the aqueous phase was extracted with $Et_2O$ and the combined organic layers were dried over $Na_2SO_4$. The organic phase was then concentrated and purified by flash chromatography (n-hexane:$CH_2Cl_2$ 9:1, later increasing the ratio to 4:1). Yield: 1.72 g (67%), red solid.

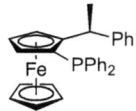

**$^1$H NMR** (700 MHz, $CD_2Cl_2$): $\delta$ = 7.59 (Ψ td, 2H, $H_{o3}$), 7.43-7.38 (m, 3H, $H_{m3/p3}$), 7.00 (t, $J$ = 7.2 Hz, 1H, $H_{p2}$), 6.98 (d, $J$ = 7.6 Hz, 2H, $H_{o1}$), 6.92-6.86 (m, 4H, $H_{m1/m2}$), 6.82 (t, $J$ = 7.2 Hz, 1H, $H_{p1}$), 6.66 (t, $J$ = 7.2 Hz, 2H, $H_{o2}$), 4.68 (s, 1H, $CH_{Cp,3}$), 4.39 (t, $J$ = 2.3 Hz, 1H, $CH_{Cp,4}$), 4.32 (qd, $J$ = 7.1 Hz, $J$ = 2.7 Hz, 1H, $CHCH_3$), 4.07 (s, 5H, $CH_{Cp'}$), 3.87 (s, 1H, $H_{Cp,5}$), 1.71 (d, $J$ = 7.1 Hz, 3H, $CH_3$).

**$^{13}$C{$^1$H} NMR** (176 MHz, $CD_2Cl_2$): $\delta$ = 147.1 (s, 1C, $C_{i1}$), 139.7 (d, $J$ = 9.1 Hz, 1C, $C_{i3}$), 138.2 (d, $J$ = 8.9 Hz, 1C, $C_{i2}$), 135.4 (d, $J$ = 21.9 Hz, 2C, $C_{o3}$), 131.7 (d, $J$ = 18.1 Hz, 2C, $C_{o2}$), 129.0 (d, $J$ = 0.6 Hz, 2C, $C_{m3}$), 127.9 (d, $J$ = 8.1 Hz, 1C, $C_{p3}$), 127.7 (s, 2C, $C_{m1}$), 127.2 (d, $J$ = 6.1 Hz, 2C, $C_{m2}$), 127.1 (d, $J$ = 1.1 Hz, 2C, $C_{o1}$), 126.8 (s, 1C, $C_{p2}$), 125.4 (s, 1C, $C_{p1}$), 100.2 (d, $J$ = 26.2 Hz, 1C, $C_{Cp,1}$), 74.8 (d, $J$ = 9.4 Hz, 1C, $C_{Cp,2}$), 71.2 (d, $J$ = 4.8 Hz, $C_{Cp,5}$), 69.6 (d, $J$ = 0.5 Hz, 5C, $C_{Cp'}$), 68.8 (s, 1C, $C_{Cp,4}$), 68.7 (d, $J$ = 4.3 Hz, 1C, $C_{Cp,3}$), 38.9 (d, $J_{P,C}$ = 9.9 Hz, 1C, $CHCH_3$), 23.2 (s, 1C, $CH_3$).

**$^{31}$P{$^1$H} NMR** (283 MHz, $CD_2Cl_2$): $\delta$ = −23.7.

**MS** (HiRes MALDI-MS): calc: [M+H$^+$]: 474.1273; found: 475.1271 (100%).
**EA**: calc. C 75.96%, H 5.74%, Fe 11.77%, P 6.53%; found C 75.74%, H 5.93% P 6.66%.
$[\alpha]_D^{20} = -270.3$ (c = 1, CHCl$_3$).
**Mp** = 184 °C.

## 5.2.23 ($S_p$)-1-Diisopropylphosphino-2-[(S)-1-phenylethyl]ferrocene (42)

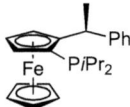

A solution of **14** (1 g, 2.71 mmol) in THF (10 mL) was cooled to −78 °C. n-BuLi (1.6 M in hexane, 1.86 mL, 2.98 mmol, 1.1 eq) was added, the color of the solution turning from orange to red. After stirring for 30 min, the cooling bath was removed and stirring was continued for another hour. Then the cooling bath was applied again and chloro di-*iso*-propylphosphine (1.31 mL, 8.13 mmol, 3 eq) was added drop wise, the solution turning from red to yellow. After stirring for 30 min, the cooling bath was removed and the reaction mixture was heated to reflux overnight. The mixture was cooled to r. t. and all volatiles were removed under HV. The residue was taken up with BH$_3$ (1 M in THF, 11 mL, 11 mmol, 4 eq). Ater 4 h, the solution was concentrated. The residue was purified by FC (n-hexane:CH$_2$Cl$_2$ 6:1). The product was dissolved in diethylamine (3 mL) and stirred at reflux for 2 h. All volatiles were evaporated under HV. Yield: 835 mg (76%), red wax.
*Due to its sensitivity, the product was not fully characterized.*
$^{31}$**P{$^1$H} NMR** (121 MHz, CD$_2$Cl$_2$): $\delta = -8.6$ (s, 1P).

## 5.2.24 ($S_p$)-1-Dicyclohexylphosphino-2-[(*S*)-1-phenylethyl]ferrocene (43)

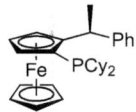

A solution of **14** (550 mg, 1.49 mmol) in THF (5 mL) was cooled to $-78\,°C$. *n*-BuLi (1.6 M in hexane, 1.0 mL, 1.6 mmol, 1.1 eq) was added, the color of the solution turning from orange to red. After stirring for 30 min, the cooling bath was removed and stirring was continued for another hour. Then the cooling bath was applied again and chloro dicyclohexylphosphine (0.99 mL, 4.47 mmol, 3 eq) was added drop wise, the solution turning from red to yellow. After stirring for 30 min, the cooling bath was removed and the reaction mixture was heated to reflux overnight. The mixture was cooled to r. t. and all volatiles were removed under HV. The residue was taken up with $BH_3$ (1 M in THF, 6 mL, 6 mmol, 4 eq). Ater 4 h, the solution was concentrated. The residue was purified by FC (*n*-hexane:$CH_2Cl_2$ 6:1). The product was dissolved in diethylamine (3 mL) and stirred at reflux for 2 h. All volatiles were evaporated under HV. Yield: 578 mg (80%), red wax.

*Due to its sensitivity, the product was not fully characterized.*
$^{31}P\{^1H\}$ **NMR** (1201 MHz, $CD_2Cl_2$): $\delta = -14.6$ (s, 1P).

## 5.2.25 ($S_p$)-1-Diphenylphosphino-2-[(*S*)-1-(3,5-dimethylphenyl)ethyl]ferrocene (44)

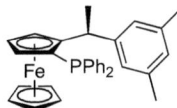

A solution of **16** (1.565 g, 9.94 mmol) in THF (30 mL) was cooled to $-78\,°C$. *n*-BuLi (1.6 M in hexane, 2.7 mL, 4.33 mmol, 1.1 eq) was added, the color of the solution turning from orange to red. After stirring for 30 min, the cooling bath was removed and stirring was continued for another hour. Then the cooling bath was applied again and chloro diphenylphosphine (2.2 mL, 12.2 mmol, 3.1 eq) was added dropwise, the solution turning from red to yellow. After stirring for 30 min, the cooling bath was removed and the reaction mixture was heated to reflux overnight, the solution turning red again. The solution was then quenched with sat. NaHCO$_3$, the aqueous phase was extracted with Et$_2$O and the combined organic layers were dried over Na$_2$SO$_4$. The organic phase was then concentrated and purified by flash chromatography (n-hexane:CH$_2$Cl$_2$ 4:1). Yield: 1.397 g (71%), red solid.

**$^1$H NMR** (400 MHz, CDCl$_3$): $\delta$ = 7.58 (m, 2H, CH$_{o3}$), 7.41 (m, 1H, CH$_{p3}$), 7.40 (m, 2H, CH$_{m3}$), 7.04 (m, 1H, CH$_{p2}$), 6.96 (m, 2H, CH$_{m2}$), 6.73 (m, 2H, CH$_{o2}$), 6.58 (s, 1H, CH$_{p1}$), 6.42 (s, 2H, CH$_{o1}$), 4.67 (dd, $J$ = 1.4, 1.0 Hz, 1H, CH$_{Cp,5}$), 4.38 (s, 1H, CH$_{Cp,4}$), 4.24 (q, $J$ = 7.2 Hz, 1H, C$H$CH$_3$), 4.09 (s, 5H, CH$_{Cp'}$), 3.85 (s, 1H, CH$_{Cp,3}$), 2.02 (s, 6H, CH$_{3,xyl}$), 1.7 (d, $J$ = 7.2 Hz, 3H, CHC$H_3$).

**$^{13}$C{$^1$H} NMR** (126 MHz, CD$_2$Cl$_2$): $\delta$ = 147.2 (s, 1C, C$_{i1}$), 137.0 (s, 2C, C$_{m1}$), 139.9 (d, $J$ = 9.2 Hz, 2C, C$_{i2}$), 138.7 (d, $J$ = 8.9 Hz, 2C, C$_{i3}$), 137.4 (s, 2C, C$_{m2}$), 135.7 (d, $J$ = 18.5 Hz, 2C, C$_{o3}$), 132.1 (d, $J$ = 18.5 Hz, 2C, C$_{o2}$), 129.3 (s, 1C, C$_{p3}$), 128.3 (d, $J$ = 8.1 Hz, 2C, C$_{m3}$), 127.4 (d, $J$ = 20.1 Hz, 2C, C$_{m2}$), 127.4 (s, 1C, C$_{p2}$), 127.9 (s, 1C, C$_{p1}$), 125.4 (s, 1C, C$_{o1}$), 100.8 (d, $J$ = 25.2 Hz, 1C, C$_{Cp,1}$ ), 75.1 (d, $J$ = 8.5 Hz, 1C, C$_{Cp,2}$), 71.6 (s, 1C, C$_{Cp,5}$), 70.0 (s, 5C, C$_{Cp'}$), 69.2 (s, 1C, C$_{Cp,3}$), 69.1 (s, 1C, C$_{Cp,4}$), 39.2 (d, $J$ = 9.2 Hz, 1C, $C$HC$_3$), 23.2 (s, 1C, CHCH$_3$), 21.2 (s, 2C, CH$_{3,xyl}$).

**$^{31}$P{$^1$H} NMR** (101 MHz, CDCl$_3$): $\delta$ = $-$23.3 (s, 1P).

**MS** (HiRes EI-MS): calc: [M$^+$]: 502.1513; found: 502.1521 (100%).

**EA**: calc. C 76.50%, H 6.22%, Fe 11.12%, P 6.17%; found C 76.31%, H 6.25%, P 6.15%.

**IR** (ATR-FT IR): $\tilde{\nu}$ = 2916 (m), 1597 (m), 1474 (m), 1449 (s), 1430 (m), 1369 (w), 1304 (w), 1231 (w), 1165 (m).

$[\alpha]_D^{20}$ = $-$251.34 (c = 1, CHCl$_3$).

**Mp** = 149 °C.

## 5.2.26 ($S_p$)-1-Diphenylphosphino-2-[($S$)-1-(1-naphthyl)ethyl]ferrocene (48)

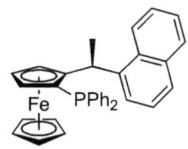

A solution of **20** (500 mg, 1.19 mmol) in THF (10 mL) was cooled to $-78\,°C$. $n$-BuLi (1.6 M in hexane, 0.85 mL, 1.36 mmol, 1.1 eq) was added, the color of the solution turning from orange to red. After stirring for 30 min, the cooling bath was removed and stirring was continued for another hour. Then the cooling bath was applied again and chloro diphenylphosphine (0.70 mL, 3.9 mmol, 3.3 eq) was added drop wise, the solution turning from red to yellow. After stirring for 30 min, the cooling bath was removed and the reaction mixture was heated to reflux overnight, and the solution turned red again. The solution was then quenched with sat. $NaHCO_3$, the aqueous phase was extracted with $Et_2O$ and the combined organic layers were dried over $Na_2SO_4$. The organic phase was then concentrated and purified by flash chromatography ($n$-hexane:$CH_2Cl_2$ 4:1). Yield: 435 mg (70%), red solid.

$^1H$ **NMR** (300 MHz, $CD_2Cl_2$): $\delta = 8.17$ (d, $J = 8.5$ Hz, 1H, $CH_{ar}$), 7.50 (d, $J = 8.1$ Hz, 1H, $CH_{ar}$), 7.38 (t, $J = 7.2$ Hz, 3H, $CH_{ar}$), 7.30–7.14 (m, 5H, $CH_{ar}$), 6.93–6.81 (m, 2H, $CH_{ar}$), 6.70 (t, $J = 7.3$ Hz, 1H, $CH_{ar}$), 6.56 (t, $J = 7.4$ Hz, 2H, $CH_{ar}$), 6.38 (t, $J = 7.5$ Hz, 2H, $CH_{ar}$), 5.06 (q, $J = 7.0$ Hz, 1H, CH), 4.65 (s, 1H, $CH_{Cp}$), 4.32 (s, 1H, $CH_{Cp}$), 4.02 (s, 5H, $CH_{Cp'}$), 3.78 (s, 1H, $CH_{Cp}$), 1.67 (d, $J = 7.0$ Hz, 3H, $CH_3$).

$^{13}C\{^1H\}$ **NMR** (175 MHz, $CD_2Cl_2$): $\delta = 143.8$ (s, 1C, $C_{ar}$), 138.7 (d, $J = 9.5$ Hz, 1C, $C_{ar}$), 138.4 (d, $J = 9.4$ Hz, 1C, $C_{ar}$), 135.3 (d, $J = 21.3$ Hz, 2C, $C_{ar}$), 133.9 (s, 1C, $C_{ar}$), 132.3 (d, $J = 19.3$ Hz, 2C, $C_{ar}$), 130.7 (s, 1C, $C_{ar}$), 129.1 (s, 1C, $C_{ar}$), 128.8 (s, 1C, $C_{ar}$), 128.3 (d, $J = 7.7$ Hz, 2C, $C_{ar}$), 127.3 (s, 1C, $C_{ar}$), 127.2 (d, $J = 6.6$ Hz, 2C, $C_{ar}$), 126.4 (s, 1C, $C_{ar}$), 125.6 (s, 1C, $C_{ar}$), 125.4 (s, 1C, $C_{ar}$), 125.2 (s, 1C, $C_{ar}$), 123.5 (s, 1C, $C_{ar}$), 100.4 (d, $J = 26.2$ Hz, 1C, $C_{Cp}$), 76.0 (d, $J = 9.1$ Hz, 1C, $C_{Cp}$), 71.8 (d, $J = 4.7$ Hz, 1C, $C_{Cp}$), 70.3 (d, $J = 4.3$ Hz, 1C, $C_{Cp}$), 70.1 (s, 5C, $C_{Cp'}$), 68.9 (s, 1C, $C_{Cp}$), 34.4 (s, 1C, CH), 22.8

(s, 1H, CH$_3$).
$^{31}$P{$^1$H} NMR (121 MHz, CD$_2$Cl$_2$): $\delta = -24.5$ (s, 1P).
MS (HiRes EI-MS): calc: [M$^+$]: 524.1356; found: 524.1363 (100%).
EA: calc. C 77.87%, H 5.57%, Fe 10.65%, P 5.91%; found C 77.91%, H 5.63%.
Mp = 133 °C.

## 5.2.27 Chloro-(($S_p$)-1-diphenylphosphino-2-[($S$)-1-phenylethyl]ferrocenyl)gold(I) (51)

In a test tube, a solution of (dimethylsulfide)gold(I)chloride (62.0 mg, 0.21 mmol, 1 eq) and ligand **40** (98.7 mg, 0.21 mmol) in CH$_2$Cl$_2$ (5 mL) was stirred for 30 min. All volatiles were removed under reduced pressure. The remaining solid was taken up in CH$_2$Cl$_2$, filtered over SiO$_2$ and the solvent was evaporated. Yield: 158.0 mg (quant), yellow solid.
$^1$H NMR (700 MHz, CD$_2$Cl$_2$): $\delta = 7.73$ (d, $J = 7.2$ Hz, 1H, H$_{o3}$), 7.71 (d, $J = 7.5$ Hz, 1H, H$_{o3}$), 7.56 ($\Psi$ td, 1H, H$_{p3}$), 7.50 ($\Psi$ td, 2H, H$_{m3}$), 7.21 ($\Psi$ td, 1H, H$_{p2}$), 7.04 ($\Psi$ td, 2H, H$_{m2}$), 7.01-6.96 (m, 4H, H$_{o1/o2}$), 6.80 ($\Psi$ t, 2H, H$_{m1}$), 6.75 ($\Psi$ t, 1H, H$_{p1}$), 5.03 (q, $J = 7.0$ Hz, 1H, C$H$CH$_3$), 4.91 (s, 1H, H$_{Cp,5}$), 4.52 (t, $J = 2.4$ Hz, 1H, H$_{Cp,4}$), 4.34 (s, 5H, H$_{Cp'}$), 3.84 (s, 1H, H$_{Cp,3}$), 1.71 (d, $J = 7.0$ Hz, 3H, CH$_3$).
$^{13}$C{$^1$H} NMR (176 MHz, CD$_2$Cl$_2$): $\delta = 145.9$ (s, 1C, C$_{i1}$), 134.8 (d, $J = 14.1$ Hz, 2C, C$_{o3}$), 132.7 (d, $J = 14.1$ Hz, 2C, C$_{o2}$), 131.7 (d, $J = 2.6$ Hz, 1C, C$_{p3}$), 130.6 (d, $J = 2.6$ Hz, 1C, C$_{p2}$), 130.0 (d, $J = 41.3$ Hz, 1C, C$_{i3}$), 129.6 (d, $J = 44.9$ Hz, 1C, C$_{i2}$), 128.7 (d, $J = 11.9$ Hz, 2C, C$_{m3}$), 128.2 (d, $J = 11.9$ Hz, 2C, C$_{m2}$), 127.9 (s, 2C, C$_{m1}$), 127.2 (s, 2C, C$_{o1}$), 125.8 (s, 1C, C$_{p1}$), 98.1 (d, $J = 15.3$ Hz, 1C, C$_{Cp,1}$), 73.6 (d, $J = 6.2$ Hz, 1C, C$_{Cp,3}$), 71.6 (d, $J_{P,C} = 8.0$ Hz, 1C, C$_{Cp,5}$), 70.9 (s, 5C, C$_{Cp'}$), 69.9 (d, $J = 8.3$ Hz, 1C, C$_{Cp,4}$), 67.1 (d, $J = 71.7$ Hz, 1C, C$_{Cp,2}$), 39.7 (d, $J = 4.0$ Hz, 1C, $C$HCH$_3$), 23.6 (s, 1C, CH$_3$).
$^{31}$P{$^1$H} NMR (122 MHz, CD$_2$Cl$_2$): $\delta = 23.4$.
MS (HiRes MALDI-MS): calc: [M$^+$]: 706.0549; found: 706.0549

(100%).
**EA**: calc. C 50.98%, H 5.85%, Au 27.87%, Cl 5.02%, Fe 7.90%, P 4.38%; found C 50.89%, H 5.88%, Cl 4.98%, P 4.59%.
$[\alpha]_D^{20} = -101.7$ (c = 1, CHCl$_3$).
**Mp** = 200 °C (decomp).

## 5.2.28 Chloro-(($S_p$)-1-diphenylphosphino-2-[($S$)-1-(1-naphthyl)ethyl]ferrocenyl)gold(I) (52)

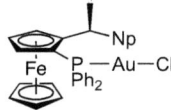

In a test tube, a solution of (dimethylsulfide)gold(I)chloride (28 mg, 0.1 mmol, 1 eq) and ligand **48** (50 mg, 0.1 mmol) in CH$_2$Cl$_2$ (2 mL) was stirred for 30 min. All volatiles were removed under reduced pressure. The remaining solid was taken up in CH$_2$Cl$_2$, filtered over SiO$_2$ and the solvent was evaporated. Yield: 72 mg (quant), yellow solid.
**$^1$H NMR** (300 MHz, CD$_2$Cl$_2$): $\delta$ = 8.20 (d, $J$ = 8.2 Hz, 1H, CH$_{ar}$), 7.56 (d, $J$ = 8.0 Hz, 1H, CH$_{ar}$), 7.32 (t, $J$ = 7.3 Hz, 3H, CH$_{ar}$), 7.28–7.10 (m, 5H, CH$_{ar}$), 6.90–6.84 (m, 2H, CH$_{ar}$), 6.74 (t, $J$ = 7.5 Hz, 1H, CH$_{ar}$), 6.55 (t, $J$ = 7.2 Hz, 2H, CH$_{ar}$), 6.33 (t, $J$ = 7.2 Hz, 2H, CH$_{ar}$), 5.01 (q, $J$ = 7.0 Hz, 1H, CH), 4.63 (s, 1H, CH$_{Cp}$), 4.27 (s, 1H, CH$_{Cp}$), 4.05 (s, 5H, CH$_{Cp'}$), 3.71 (s, 1H, CH$_{Cp}$), 1.70 (d, $J$ = 7.0 Hz, 3H, CH$_3$).
**$^{13}$C{$^1$H} NMR** (75 MHz, CD$_2$Cl$_2$): $\delta$ = 143.1 (s, 1C, C$_{ar}$), 138.7 (d, $J$ = 9.5 Hz, 1C, C$_{ar}$), 138.3 (d, $J$ = 9.4 Hz, 1C, C$_{ar}$), 135.1 (d, $J$ = 21.3 Hz, 2C, C$_{ar}$), 133.1 (s, 1C, C$_{ar}$), 132.8 (d, $J$ = 19.3 Hz, 2C, C$_{ar}$), 130.1 (s, 1C, C$_{ar}$), 129.0 (s, 1C, C$_{ar}$), 128.5 (s, 1C, C$_{ar}$), 128.1 (d, $J$ = 7.7 Hz, 2C, C$_{ar}$), 127.9 (s, 1C, C$_{ar}$), 127.0 (d, $J$ = 6.6 Hz, 2C, C$_{ar}$), 126.7 (s, 1C, C$_{ar}$), 125.2 (s, 1C, C$_{ar}$), 125.1 (s, 1C, C$_{ar}$), 125.0 (s, 1C, C$_{ar}$), 123.1 (s, 1C, C$_{ar}$), 98.9 (d, $J$ = 26.2 Hz, 1C, C$_{Cp}$), 76.9 (d, $J$ = 9.1 Hz, 1C, C$_{Cp}$), 71.0 (d, $J$ = 4.7 Hz, 1C, C$_{Cp}$), 70.1 (d, $J$ = 4.3 Hz, 1C, C$_{Cp}$), 70.0 (s, 5C, C$_{Cp'}$), 68.2 (s, 1C, C$_{Cp}$), 34.9 (s, 1C, CH), 21.7 (s, 1H, CH$_3$).
**$^{31}$P{$^1$H} NMR** (121 MHz, CD$_2$Cl$_2$): $\delta$ = −27.3 (s, 1P).

**MS** (HiRes EI-MS): calc: [M$^+$]: 756.0710; found: 756.0718 (100%).
**EA**: calc. C 53.96%, H 3.86%, Au 26.03%, Cl 4.68%, Fe 7.38%, P 4.09%; found C 53.91%, H 3.80%.
**Mp** = 230 °C (decomp).

### 5.2.29 Methyl-(($S_p$)-1-diphenylphosphino-2-[($S$)-1-phenylethyl]ferrocenyl)gold(I) (55)

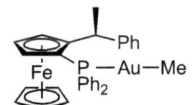

A solution of iodomethane (0.15 mL, 2.41 mmol, 8.6 eq) in Et$_2$O (2 mL) was added dropwise to magnesium turnings (81 mg, 3.33 mmol, 11.9 eq). After the addition was complete, the mixture was heated to reflux for 30 min. The Grignard reagent was then slowly added to a solution of complex **51** (198 mg, 0.28 mmol) in Et$_2$O (2 mL) at −15 °C (NaCl/ice-bath). The solution was stirred at r. t. for 1 h then heated to reflux for 2 h. The mixture was then poured into ice-cold diluted H$_2$SO$_4$ (80 mL, 0.5%). The phases were separated, the aqueous phase was extracted with Et$_2$O and the combined organic phases were dried over Na$_2$SO$_4$. Evaporation of the solvent gave an orange powder which was recrystallized from toluene/pentane at −18 °C. Yield: 155 mg (81%), orange crystals.

*As the product decomposed in CD$_2$Cl$_2$, only selected signals are given.*
**$^1$H NMR** (300 MHz, CD$_2$Cl$_2$): $\delta$ = 0.55 (d, $J$ = 8.0 Hz, 3H, AuMe).
**$^{13}$C{$^1$H} NMR** (75 MHz, CD$_2$Cl$_2$): $\delta$ = 6.1 (d, $J$ = 96.6 Hz, 1C, AuMe).
**$^{31}$P{$^1$H} NMR** (121 MHz, CD$_2$Cl$_2$): $\delta$ = 38 (s, 1P).
**MS** (HiRes EI-MALDI): calc: [M$^+$]: 686.1100; found: 686.1102 (100%).

## 5.2.30 Bis(($S_p$)-1-diphenylphosphino-2-[($S$)-1-phenylethyl]ferrocenyl)gold(I) hexafluorophosphate (56)

A solution of dimethylsulfide (1 mL) and (dimethylsulfide)gold(I) chloride (93.6 mg, 0.32 mmol, 1 eq) in $CH_2Cl_2$ (10 mL) was added to a solution of silver hexafluorophosphate (80.1 mg, 0.32 mmol) and stirred for 1.5 h. The solution was then filtered over Celite®, and concentrated under HV. The residue was covered with a layer of $Et_2O$ and was stored at −18 °C overnight to precipitate a white solid. The solvents were evaporated and the white solid was dissolved in $CH_2Cl_2$ followed by addition of the phosphine **40** (150.6 mg, 0.32 mmol, 1 eq). After stirring for 5 h, the mixture was purified by flash chromatography first using $CH_2Cl_2$ to elute formed **51**, then EtOAc was used to elute the product. Yield: 134.4 g (68%), yellow solid.

**$^1$H NMR** (300 MHz, $CD_2Cl_2$): $\delta$ = 7.85–7.72 (m, 4H, $CH_{ar}$), 7.70–7.64 (m, 6H, $CH_{ar}$), 7.36–7.30 (m, 2H, $CH_{ar}$), 7.16-7.10 (m, 4H, $CH_{ar}$), 7.05–7.01 (m, 4H, $CH_{ar}$), 6.85–6.69 (m, 10H, $CH_{ar}$), 5.04 (s, 2H, $CH_{Cp}$), 4.85 (q, $J$ = 6.9 Hz, 2H, $CH$(Ph)$CH_3$), 4.77 (Ψ t, 2H, $CH_{Cp}$), 4.41 (s, 10H, $CH_{Cp'}$), 4.06 (s, 2H, $CH_{Cp}$), 1.73 (d, $J$ = 7.2 Hz, 6H, $CH_3$).

**$^{31}$P{$^1$H} NMR** (122 MHz, $CD_2Cl_2$): $\delta$ = 35.7 (br, 2P, Fc-PPh2-Au), -144.3 (sept, $J_{P,F}$ = 710 Hz, 1P, $PF_6$).

## 5.2.31 Dichloro-(($S_p$)-1-$\kappa$P-diphenylphosphino-2-[($R$)-1-($\eta^6$-phenyl)ethyl]ferrocenyl)-ruthenium(II) (57)

In a *Young*-Schlenk, a solution of [Ru($\eta^6$-$p$-cymene)Cl$_2$]$_2$ (92.3 mg, 0.15 mmol, 0.5 eq) and ligand 40 (142 mg, 0.3 mmol) in toluene (8 mL) was stirred at r. t. for 1 h. The solution was heated to reflux, the 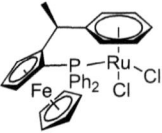 flask closed and stirred for an additional 15 h. The mixture was let cool down to r. t., filtered over Celite® and all volatiles were removed *in vacuo*. Yield: 173 mg (97%), brown solid.

$^1$**H NMR** (500 MHz, CD$_2$Cl$_2$): $\delta$ = 7.91 (m, 2H, C$H_{m2}$), 7.57 (m, 2H, C$H_{m3}$), 7.49 (m, 1H, C$H_{p2}$), 7.42 (m, 2H, C$H_{o2}$), 7.41 (m, 1H, C$H_{p3}$), 7.33 (m, 2H, C$H_{o3}$), 6.37 (m, 1H, C$H_{p1}$), 6.12 (t, $J$ = 6 Hz, 1H, C$H_{m1}$), 5.97 (d, $J$ = 5.4 Hz, 1H, C$H_{o1}$), 5.31 (t, $J$ = 5.6 Hz, 1H, C$H_{m1}$), 4.79 (m, 1H, C$H_{Cp,3}$), 4.72 (t, $J$ = 5 Hz, 1H, C$H_{o1}$), 4.69 (t, $J$ = 2.6 Hz, 1H, C$H_{Cp,5}$), 4.33 (s, 5H, C$H_{Cp'}$), 4.16 (m, 1H, C$H_{Cp,4}$), 4.10 (q, $J$ = 7.2 Hz, 1H, C$H$CH$_3$), 1.35 (d, $J$ = 7.3 Hz, 3H, CH$_3$).

$^{13}$**C{$^1$H} NMR** (126 MHz, CD$_2$Cl$_2$): $\delta$ = 138.1 (d, $J_{P,C}$ = 75.5 Hz, 2C, C$_{i3}$), 135.8 (d, $J_{P,C}$ = 9.6 Hz, 2C, C$_{m2}$), 133.4 (d, $J_{P,C}$ = 8.9 Hz, 2C, C$_{m3}$), 131.2 (d, $J_{P,C}$ = 74.2 Hz, 1C, C$_{i2}$), 130.8 (d, $J_{P,C}$ = 3.1 Hz, 1C, C$_{p2}$), 130.4 (d, $J_{P,C}$ = 2.3 Hz, 2C, C$_{o2}$), 128.0 (d, $J_{P,C}$ = 10.0 Hz, 2C, C$_{o3}$), 127.6 (2, $J_{P,C}$ = 10.8 Hz, 1C, C$_{p3}$), 99.7 (s, 1C, C$_{Cp,1}$), 98.4 (d, $J_{P,C}$ = 11.2 Hz, 1C, C$_{p1}$), 97.4 (d, $J_{P,C}$ = 19.7 Hz, 1C, C$_{i1}$), 95.3 (d, $J_{P,C}$ = 7.3 Hz, 1C, C$_{m1}$), 86.7 (d, $J$ = 2.3 Hz, 1C, C$_{m2}$), 85.3 (s, 1C, C$_{o1}$), 84.7 (s, 1C, C$_{o2}$), 72.0 (d, $J_{P,C}$ = 5.4 Hz, 1C, C$_{Cp,3}$), 71.5 (s, 5C, C$_{Cp'}$), 71.4 (s, 1C, C$_{Cp,4}$), 71.3 (d, $J_{P,C}$ = 8.1 Hz, 1C, C$_{Cp,5}$), 70.8 (d, $J_{P,C}$ = 54 Hz, 1C, C$_{Cp,2}$), 34.4 (s, 1C, $C$HCH$_3$), 24.5 (s, 1C, CH$_3$).

$^{31}$**P{$^1$H} NMR** (203 MHz, CD$_2$Cl$_2$): $\delta$ = 22.6 (s, 1P).

**MS** (HiRes MALDI-MS): calc: [(M−$Cl$)$^+$]: 610.9932; found: 610.9937 (100%).

**EA**: calc. C 55.75%, H 4.21%, Cl 10.97%, Fe 8.64%, P 4.79%, Ru 15.64%; found C 55.80%, H 4.25%, Cl 10.85%, P 4.69%.

**IR** (ATR-FT IR): $\tilde{\nu}$ = 3069 (m), 3044 (w), 2990 (w), 2965 (m), 2932 (w), 1481 (m), 1463 (s), 1448 (s), 1432 (m), 1406 (m), 1260 (m), 1184 (w), 1157 (m).
$[\alpha]_D^{20} = -45.41$ (c = 1, CHCl$_3$).
**Mp** = 279 °C.

### 5.2.32 Dichloro-((*S$_p$*)-1-κP-diisopropylphosphino-2-[(*R*)-1-($\eta^6$-phenyl)ethyl]ferrocenyl)-ruthenium(II) (58)

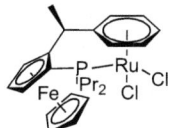

In a *Young*-Schlenk, a solution of [Ru($\eta^6$-p-cymene)Cl$_2$]$_2$ (326 mg, 0.53 mmol, 0.5 eq) and ligand **42** (433 mg, 1.06 mmol) in toluene (10 mL) was stirred at r. t. for 1 h. The solution was heated to reflux, the flask closed and stirred for an additional 3 h. The mixture was let cool down to r. t., filtered over Celite® and all volatiles were removed *in vacuo*. The crude product was purified by recrystallization from methanol and removal of the solfent under HV. Yield: 421 mg (68%), red crystals.

**$^1$H NMR** (300 MHz, CD$_2$Cl$_2$): $\delta$ = 5.95 (t, *J* = 5.8 Hz, 1H, CH$_{ar}$), 5.79 (t, *J* = 5.8 Hz, 1H, CH$_{ar}$), 5.57 (d, *J* = 5.3 Hz, 1H, CH$_{ar}$), 5.42 (t, *J* = 5.6 Hz, 1H, CH$_{ar}$), 4.68 (s, 1H, CH$_{Cp}$), 4.52 (s, 2H, CH$_{Cp}$+CH$_{ar}$), 4.44 (s, 1H, CH$_{Cp}$), 4.21 (s, 5H, CH$_{Cp'}$), 3.69 (q, *J* = 7.0 Hz, 1H, C*H*(Ph)CH$_3$), 2.88–2.72 (m, 1H CH$_{iPrB}$), 2.51–2.33 (m, 1H CH$_{iPrA}$), 1.50 (dd, *J* = 15.5, 7.7 Hz, 3H, CH$_{3,iPrB}$),1.45 (dd, *J* = 16.9, 8.2 Hz, 3H, CH$_{3,iPrB}$),1.43 (d, *J* = 7.9 Hz, 3H, CH$_3$), 1.01 (ddd, *J* = 24.4, 16.2, 7.3 Hz, 6H, CH$_{3,iPrA}$).

**$^{13}$C{$^1$H} NMR** (101 MHz, CDCl$_3$): $\delta$ = 105.1 (s, 1C, C$_{ar}$), 97.5 (d, *J* = 10.4 Hz, 1C, C$_{ar}$), 96.2 (d, *J* = 5.6 Hz, 1C, C$_{ar}$), 95.5 (d, *J* = 14.8 Hz, 1C, C$_{Cp}$), 91.4 (d, *J* = 2.2 Hz, 1C, C$_{ar}$), 79.0 (s, 1C, C$_{ar}$), 78.7 (s, 1C, C$_{ar}$), 73.4 (d, *J* = 2.1 Hz, 1C, C$_{Cp}$), 71.3 (s, 5C, C$_{Cp'}$), 70.0 (d, *J* = 7.2 Hz, 1C, C$_{Cp}$), 69.4 (d, *J* = 5.2 Hz, 1C, C$_{Cp}$), 35.9 (s, 1C, *c*HCH$_3$), 28.0 (d, *J* = 25.9 Hz, 1C, C$_{iPrA}$), 26.2 (d, *J* = 25.3 Hz,

1C, $C_{iPrB}$), 22.4 (s, 1C, $CH_3$), 19.2 (s, 1C, $C_{iPrB}$), 18.0 (d, $J = 4.7$ Hz, 2C, $C_{iPrA}$), 17.7 (d, $J = 4.0$ Hz, 1C, $C_{iPrB}$).
**MS** (HiRes EI-MS): calc: [M$^+$]: 577.9933; found: 577.9928 (100%).
**EA**: calc. C 49.85%, H 5.40%, Cl 12.26%, Fe 9.66%, P 5.36%, Ru 17.48%; found C 49.87%, H 5.42%, Cl 12.20%.
**Mp** = 186 °C.

## 5.2.33 Dichloro-(($S_p$)-1-$\kappa$P-diphenylphosphino-2-[($R$)-1-(3,5-dimethyl-($\eta^6$-phenyl))ethyl]-ferrocenyl)ruthenium(II) (60)

In a *Young*-Schlenk, a solution of [Ru($\eta^6$-p-cymene)Cl$_2$]$_2$ (189 mg, 0.31 mmol, 0.5 eq) and ligand **44** (310 mg, 0.62 mmol) in toluene (12 mL) was stirred at r.t. for 1 h. The solution was heated to

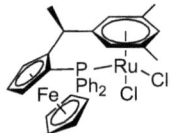

reflux, the flask closed and stirred for an additional 15 h. The mixture was let cool down to r.t., filtered over Celite® and all volatiles were removed *in vacuo*. The crude product was purified by FC (n-hexane:EtOAc 1:2). Yield: 351 mg (85%), brown solid.
**$^1$H NMR** (500 MHz, CD$_2$Cl$_2$): $\delta$ = 7.98 (m, 2H, $CH_{m2}$), 7.64 (m, 2H, $CH_{m3}$), 7.47 (m, 1H, $CH_{p2}$), 7.41 (m, 1H, $CH_{p3}$), 7.39 (m, 2H, $CH_{o2}$), 7.34 (m, 2H, $CH_{o3}$), 5.73 (m, 1H, $CH_{p1}$), 5.07 (m, 1H, $CH_{o1}$), 4.73 (m, 1H, $CH_{Cp,3}$), 4.64 (m, 1H, $CH_{Cp,4}$), 4.44 (s, 5H, $CH_{Cp'}$), 4.16 (m, 1H, $CH_{o1}$), 4.02 (m, 1H, $CH_{Cp,5}$), 4.02 (q, 1H, $J = 7.3$ Hz, $CHC_3$), 2.37 (s, 3H, $CH_{3,Xyl}$), 1.27 (s, 3H, $CH_{3,Xyl}$), 1.27 (d, $J = 7.3$ Hz, 3H, $CHCH_3$).
**$^{13}$C{$^1$H} NMR** (126 MHz, CD$_2$Cl$_2$): $\delta$ = 139.3 (d, $J = 49.1$ Hz, 2C, $C_{o3}$), 136.0 (d, $J = 9.2$ Hz, 2C, $C_{m2}$), 133.5 (d, $J = 8.9$ Hz, 2C, $C_{m3}$), 131.3 (d, $J = 49.1$ Hz, 1C, $C_{i2}$), 130.6 (d, $J = 26.5$ Hz, 1C, $C_{p2}$), 130.5 (2, $J = 26$ Hz, 1C, $C_{p3}$), 128.0 (d, $J = 10.0$ Hz, 2C, $C_{o3}$), 127.4 (d, $J = 10.4$ Hz, 2C, $C_{o2}$), 112.0 (d, $J = 7.7$ Hz, 1C, $C_{m1}$), 102.2 (s, 1C, $C_{i1}$), 95.5 (d, $J = 7.3$ Hz, 1C, $C_{m1}$), 97.2 (d, $J = 18.9$ Hz, 1C, $C_{Cp,1}$), 95.3 (d, $J_{P,C} = 12.7$ Hz, 1C, $C_{p1}$), 83.3 (s, 1C, $C_{o1}$), 80.5 (s, 1C, $C_{o1}$), 71.9 (d, $J_{P,C} = 71.7$ Hz, 1C, $C_{Cp,2}$), 71.8 (d, $J = 5.4$ Hz, 1C, $C_{Cp,4}$),

71.3 (s, 5C, $C_{Cp'}$), 71.2 (m, 1C, $C_{Cp,3}$), 71.1 (m, 1C, $C_{Cp,5}$), 34.3 (s, 1C, $C$HCH$_3$), 24.4 (s, 1C, CH$C$H$_3$), 18.6 (s, 1C, CH$_{3,Xyl}$), 17.9 (s, 1C, CH$_{3,Xyl}$).
$^{31}$P{$^1$H} NMR (203 MHz, CD$_2$Cl$_2$): $\delta = 22.1$ (s, 1P)
MS (HiRes MALDI-MS): calc: [(M−Cl)$^+$]: 639.0245; found: 639.0253 (100%).
EA: calc. C 56.91%, H 4.63%, Cl 10.51%, Fe 8.28%, P 4.59%, Ru 14.99%; found C 57.13%, H 4.84%, P 4.42%.
IR (ATR-FT IR): $\tilde{\nu} = 3057$ (w), 2959 (w), 2925 (w), 1535 (w), 1482 (w), 1433 (w), 1374 (w), 1261 (m), 1190 (w), 1158 (m), 1089 (m).
$[\alpha]_D^{20} = 13.31$ (c = 1, CHCl$_3$).
Mp = 275 °C.

## 5.2.34 Dichloro-(($S_p$)-1-$\kappa$P-diphenylphosphino-2-(($S$)-1-phenylethyl)ferrocenyl)-[(1,2,3,4,5-$\eta$)-1,2,3,4,5-pentamethyl-cyclopentadienyl]iridium(III) (64)

Phosphine **69** (20.5 mg, 0.04 mmol, 2 eq) and [Ir(Cp*)Cl$_2$]$_2$ (16.3 mg, 0.02 mmol) were dissolved in CH$_2$Cl$_2$ (0.5 mL) and stirred for 30 min. The solution was filtered through Celite® and the solvent was removed *in vacuo*. Yield: 35 mg (93%), red crystals.
$^1$H NMR (300 MHz, CD$_2$Cl$_2$): $\delta = 8.35$ (m, 2H, CH$_{ar}$), 7.47 (s, 3H, CH$_{ar}$), 7.03 (m, 2H, CH$_{ar}$), 6.87 (m, 4H, CH$_{ar}$), 6.71 (m, 4H, CH$_{ar}$), 5.10 (s, 1H, CH$_{Cp}$), 4.53 (s, 1H, CH$_{Cp}$), 4.48 (s, 1H, CH$_{Cp}$), 3.95 (s, 5H, CH$_{Cp'}$), 3.28 (q, $J = 7.2$ Hz, 1H, C$H$CH$_3$), 1.24 (q, $J = 7.2$ Hz, 1H, CHC$H_3$), 1.05 (s, 15H, CH$_{3,Cp*}$).
$^{13}$C{$^1$H} NMR (75 MHz, CD$_2$Cl$_2$): $\delta = 145.9$ (s, 1C, C$_{ar}$), 134.8 (s, 2C, C$_{ar}$), 132.7 (s, 2C, C$_{ar}$), 131.7 (s, 1C, C$_{ar}$), 130.6 (s, 1C, C$_{ar}$), 130.0 (s, 1C, C$_{ar}$), 129.6 (s, 1C, C$_{ar}$), 128.7 (s, 2C, C$_{ar}$), 128.2 (s, 2C, C$_{ar}$), 127.9 (s, 2C, C$_{ar}$), 127.2 (s, 2C, C$_{ar}$), 125.8 (s, 1C, C$_{ar}$), 98.1 (s,

1C, $C_{Cp}$), 92.7 (s, 5C, $C_{q,Cp*}$), 73.6 (s, 1C, $C_{Cp}$), 71.6 (s, $J = 8.0$ Hz, 1C, $C_{Cp}$), 71.5 (s, 5C, $C_{Cp'}$), 69.9 (s, 1C, $C_{Cp}$), 67.1 (s, 1C, $C_{Cp}$), 37.2 (s, 1C, $CHCH_3$), 28.8 (s, 1C, $CH_3$) 8.1 (s, 5C, $CH_{3,Cp*}$).
$^{31}$P{$^1$H} NMR (121 MHz, $CD_2Cl_2$): $\delta = -8.5$ (s, 1P).
MS (HiRes EI-MS): calc: [M$^+$]: 872.1380; found: 872.1387 (100%).
EA: calc. C 55.05%, H 4.85%, Cl 8.12%, Fe 6.40%, Ir 22.03%, P 3.55%; found C 55.11%, H 4.79%.
Mp = 216 °C.

## 5.2.35 Dichloro-(($S_p$)-1-$\kappa$P-diphenylphosphino-2-(($S$)-1-phenylethyl)ferrocenyl)[(1,2,3,4,5-$\eta$)-1,2,3,4,5-pentamethyl-cyclopentadienyl]rhodium(III) (65)

Phosphine **69** (20 mg, 0.04 mmol, 2 eq) and [Rh(Cp*)Cl$_2$]$_2$ (13 mg, 0.02 mmol) were dissolved in CH$_2$Cl$_2$ (0.5 mL) and stirred for 30 min. The solution was filtered through Celite® and the solvent was removed *in vacuo*. Yield: 29 mg (88%), red crystals.

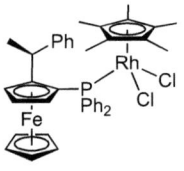

$^1$H NMR (250 MHz, CDCl$_3$): $\delta = 8.37$ (m, 2H, $CH_{ar}$), 7.53 (s, 3H, $CH_{ar}$), 7.02 (m, 2H, $CH_{ar}$), 6.89 (m, 4H, $CH_{ar}$), 6.65 (m, 4H, $CH_{ar}$), 5.13 (s, 1H, $CH_{Cp}$), 4.62 (s, 1H, $CH_{Cp}$), 4.42 (s, 1H, $CH_{Cp}$), 3.94 (s, 5H, $CH_{Cp'}$), 3.25 (q, $J = 7.2$ Hz, 1H, $CHCH_3$), 1.24 (q, $J = 7.2$ Hz, 1H, $CHCH_3$), 1.01 (s, 15H, $CH_{3,Cp*}$).
$^{13}$C{$^1$H} NMR (62 MHz, CDCl$_3$): $\delta = 146.3$ (s, 1C, $C_{ar}$), 132.3 (s, 2C, $C_{ar}$), 135.2 (s, 2C, $C_{ar}$), 130.9 (s, 1C, $C_{ar}$), 130.3 (s, 1C, $C_{ar}$), 130.2 (s, 1C, $C_{ar}$), 129.5 (s, 1C, $C_{ar}$), 128.6 (s, 2C, $C_{ar}$), 128.4 (s, 2C, $C_{ar}$), 127.8 (s, 2C, $C_{ar}$), 127.1 (s, 2C, $C_{ar}$), 125.9 (s, 1C, $C_{ar}$), 98.2 (s, 1C, $C_{Cp}$), 96.3 (s, 5C, $C_{q,Cp*}$), 73.8 (s, 1C, $C_{Cp}$), 71.4 (s, $J = 8.0$ Hz, 1C, $C_{Cp}$), 71.1 (s, 5C, $C_{Cp'}$), 69.5 (s, 1C, $C_{Cp}$), 67.0 (s, 1C, $C_{Cp}$), 37.5 (s, 1C, $CHCH_3$), 28.2 (s, 1C, $CH_3$) 9.0 (s, 5C, $CH_{3,Cp*}$).
$^{31}$P{$^1$H} NMR (81 MHz, $CD_2Cl_2$): $\delta = 22.8$ (d, $J = 144$ Hz, 1P).

**MS** (HiRes EI-MS): calc: [M$^+$]: 782.0806; found: 782.0817 (100%).
**EA**: calc. C 61.33%, H 5.40%, Cl 9.05%, Fe 7.13%, P 3.95%, Rh 13.14%; found C 61.40%, H 5.43%.
**Mp** = 206 °C.

## 5.2.36 ($S_p$)-1-Diphenylphosphino-2-[($R$)-1-(1$H$-indol-1-yl)ethyl]ferrocene (66)

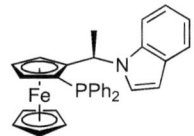

A solution of ppfa (250 mg, 0.56 mmol) and indole (133 mg, 1.12 mmol, 2 eq) in degassed acetic acid (3 mL) was stirred at reflux for 5 h. The misture was cooled to r. t. and all volatiles were removed under HV. The crude product was purified by FC (cyclohexane:EtOAc:NEt$_3$ 200:30:1). Yield: 164 mg (56%), orange solid.

$^1$**H NMR** (300 MHz, CDCl$_3$): $\delta$ = 7.71–6.93 (m, 11H, CH$_{ar}$), 6.73 (b, 2H, CH$_{ar}$), 6.59 (b, 3H, CH$_{ar}$), 4.80 (s, 1H, CH$_{Cp}$), 4.64 (s, 1H, CH), 4.37 (s, 1H, CH$_{Cp}$), 4.15 (s, 5H, CH$_{Cp'}$), 3.80 (s, 1H, CH$_{Cp}$), 1.89 (s, 3H, CH$_3$).
$^{13}$**C{$^1$H} NMR** (75 MHz, CDCl$_3$): $\delta$ = 138.9 (d, $J$ = 7.1 Hz, 1C, C$_{ar}$), 138.0 (d, $J$ = 8.3 Hz, 1C, C$_{ar}$), 136.6 (s, 1C, C$_{ar}$), 135.3 (d, $J$ = 21.1 Hz, 2C, C$_{ar}$), 131.7 (d, $J$ = 18.3 Hz, 2C, C$_{ar}$), 128.9 (s, 1C, C$_{ar}$), 127.9 (d, $J$ = 7.8 Hz, 2C), 126.9 (d, $J$ = 6.4 Hz, 2C), 126.6 (s, 1C, C$_{ar}$), 125.7 (s, 1C, C$_{ar}$), 121.9 (d, $J$ = 2.4 Hz, 2C, C$_{ar}$), 121.6 (s, 1C, C$_{ar}$), 121.2 (s, 1C, C$_{ar}$), 119.8 (s, 1C, C$_{ar}$), 118.7 (s, 2C, C$_{ar}$), 110.9 (s, 1C, C$_{ar}$), 99.6 (d, $J$ = 24.0 Hz, 2C, C$_{q,Cp}$), 74.8 (d, $J$ = 7.4 Hz, 1C, C$_{q,Cp}$), 71.3 (d, $J$ = 4.6 Hz, 2C, CH$_{Cp}$), 70.3 (s, 1C, CH$_{Cp}$), 69.7 (s, 5C, CH$_{Cp'}$), 69.3 (d, $J$ = 4.2 Hz, 1C, CH$_{Cp}$), 68.3 (s, 2C, CH$_{Cp}$), 30.9 (d, $J$ = 9.3 Hz, 1C, CH$_3$), 21.5 (s, 1C, CH$_3$).
$^{31}$**P{$^1$H} NMR** (121 MHz, CD$_2$Cl$_2$): $\delta$ = −23.2 (s, 1P).
**MS** (HiRes EI-MS): calc: [M$^+$]: 513.1309; found: 513.1318 (100%).
**EA**: calc. C 74.86%, H 5.50%, Fe 10.88%, N 2.73%, P 6.03%; found C 74.91%, H 5.47%, N 2.71%.

**Mp** = 212 °C.

## 5.3 Catalyses

### 5.3.1 Michael Addition of Diethyl Malonate to Nitrostyrene

*Catalysis was carried out according to Pedrosa and co-workers [137]*
Diethyl malonate (90 µL, 0.6 mmol, 2 eq) was added to a solution of *trans*-$\beta$-nitrostyrene (45.6 mg, 0.3 mmol) and urea **33** (0.1 eq) at r. t.. The reaction was monitored by TLC (*n*-hexane:EtOAc 20:1). As no product formation could be detected after 24 h, the temperature was raised stepwise to reflux (+20 °C after 24 h). No product was formed.

### 5.3.2 In$^0$-mediated Allylation of Hydrazones

*Catalyses were carried out according to Jacobsen and co-workers.[138]*
(*E*)-*N*'-benzylidenebenzohydrazone (56 mg, 0.25 mmol), indium powder (50 mg, 0.44 mmol, 1.74 eq) and the catalyst (25 µmol, 0.1 eq) were suspended in toluene (2.5 mL) and cooled to −78 °C. Allyl bromide (57 µL, 0.66 mmol, 2.6 eq) was added and the mixtures was allowed to warm to −10 °C and stirred for 24 h. The reaction was quenched with HCl (1 M) and the mixture was neutralized with NaOH (0.6 M). The product was extracted with EtOAc, the organic phases were dried over Na$_2$SO$_4$ and the solvents were removed under reduced pressure. The crude product was purified three times by flash chromatography: 2 × SiO$_2$, toluene:EtOAC 5:1, 1% NEt$_3$; 1 × Alox B, *n*-hexane:EtOAc 5:1).
**HPLC**: AM, 6% *i*-PrOH/*n*-hexane, 1.0 mL/ min.  $t_r$ (major) = 18.4 min, $t_r$ (minor) = 21.6 min.

| entry | catalyst | crude yield /[%] | isolated yield /[%] | $[\alpha]_D^{20} = /[°]^a$ | er[b] |
|---|---|---|---|---|---|
| 1 | 33 | quant | 44 | +39 | 68:32 |
| 2 | 32 | quant[c] | 46 | +45 | 66:34 |

[a]CHCl$_3$, values normalized for c = 1. [b] Determined by HPLC. [c]conversion was not complete after 24 h therefore the reaction mixture was allowed to warm to r. t. over 3 h.

### 5.3.3 Au$^I$-Catalyzed Intramolecular Hydroamination

#### 5.3.3.1 Halide Abstraction Screening

Chloro complex **51** (ca. 10 mg, 14 μmol) and the halogen scavenger (1 eq) were put in an NMR tube. Solvent was added and conversion was monitored by $^{31}$P{$^1$H} NMR.

#### 5.3.3.2 Catalysis

*Catalyses were attempted according to Widenhoefer and coworkers.[241]*
5 mol-% catalyst and (where applicable) 5 mol-% halogen scavenger were dissolved in dioxane (2 mL). The substrate was added and the mixture was degassed by a freeze-pump-thaw cycle. The solution was then heated to 60 °C for 18 h. The reaction mixture was purified by flash chromotography (*n*-hexane:EtOAc 20:1).

### 5.3.4 Au$^I$-Catalyzed Hydroarylation of Styrene

To a suspension of complex **55** in toluene (2 mL), 1-methylindol (40 μL, 0.31 mmol), 4-methylstyrene (46 μL, 0.35 mmol, 1.1 eq) and the acid was added. The mixture was stirred for 3 h at 85 °C.

| entry | mol-% cat | acid | observation |
|---|---|---|---|
| 1 | 2 | $H_2SO_4$ (20 mol-%) | rosy solution, red precipitate, no product |
| 2 | 5 | $H_2SO_4$ (20 mol-%) | rosy solution, red precipitate, no product |
| 3 | 5 | $H_2SO_4$ (50 mol-%) | rosy solution, red precipitate, no product |
| 3 | 5 | $H_2SO_4$ (100 mol-%) | rosy solution, red precipitate, no product |
| 4 | 5 | $HBF_4 \cdot Et_2O$ (10 mol%) | strong gas evolution upon addition, dark precipitate |

## 5.3.5 Ru$^{II}$-Catalyzed Transfer Hydrogenation

A solution of Meerwein salt (ca. 5% in $CH_2Cl_2$) was added to the ruthenium complex (5 mol-%). After 3 h, $i$-PrOH (1 mL), base and acetophenone (18 µL, 15.5 mmol) was added. The mixture was heated to the given temperature for 20 h. Sat. aq. $NH_4Cl$ (1 mL) was added and the mixture was extracted with $Et_2O$. The organic phase was filtered through a short plug of silica and concentrated *in vacuo*.

**GC**: $\beta$-DEX, 80 °C. $t_r$ ($R$) = 105 min, $t_r$ ($S$) = 120 min. Enantiomeric excess below 5% is considered racemic.

| entry | cat | base | X$^-$ | T/[°C] | yield/[%] | er | comment |
|---|---|---|---|---|---|---|---|
| 1 | 57 | - | - | 60 | <10 | rac | |
| 2 | 57 | - | - | 80 | 39 | rac | |
| 3 | 57 | KO$t$-Bu | PF$_6$ | 60 | >99 | rac | |
| 4 | 57 | KO$t$-Bu | - | 60 | >99 | rac | |
| 5 | 57 | KO$t$-Bu | PF$_6$ | 60 | >99 | rac | 1 mol-% **57**, PF$_6$ |
| 6 | 57 | KO$t$-Bu | - | 60 | >99 | rac | 1 mol-% **57** |
| 7 | 57 | KO$t$-Bu | - | 40 | >99 | rac | 1 mol-% **57** |
| 8 | 57 | - | PF$_6$ | 60 | 6 | rac | |
| 9 | 57 | - | PF$_6$ | 60 | 15 | rac | |
| 10 | 57 | - | PF$_6$ | 80 | 30 | rac | |
| 11 | 57 | - | PF$_6$ | 80 | 62 | rac | NEt$_3$/HCOOH$^b$ 2:5 |
| 12 | 57 | NEt$_3$ | - | 40 | 51 | rac | |
| 13 | 57 | NEt$_3$ | - | 60 | 87 | rac | |
| 14 | 57 | NEt$_3$ | - | 60 | 94 | rac | |
| 15 | 57 | NEt$_3$ | - | 80 | 90 | rac | |
| 16 | 57 | NEt$_3$ | - | 80 | 95 | rac | |
| 17 | 57 | dbu | - | 80 | 93 | rac | |
| 18 | 57 | dipea | - | 80 | 72 | rac | |
| 19 | 57 | pyridine | - | 80 | 2 | rac | |
| 20 | 57 | NH$_3$ | - | 80 | 35 | rac | |
| 21 | 57 | NEt$_3$ | BF$_4$ | 60 | <5 | n.d. | |
| 22 | 57 | NEt$_3$ | BF$_4$ | 80 | 5 | n.d. | |
| 23 | 57 | NEt$_3$ | PF$_6$ | 60 | 10 | 57:43 | |

| entry | cat | base | X$^-$ | T /[°C] | yield [%] | er | comment |
|---|---|---|---|---|---|---|---|
| 24 | **57** | NEt$_3$ | PF$_6$ | 80 | 25 | 61:39 | |
| 25 | **57** | NEt$_3$ | SbF$_6$ | 80 | 50 | 67:33 | |
| 26 | **57** | NEt$_3$ | PF$_6$ | 80 | 25 | 62:38 | |
| 27 | **57** | NEt$_3$ | PF$_6$ | 80 | 25 | 62:38 | |
| 28 | **57** | NEt$_3$ | PF$_6$ | 80 | 18 | 61:39 | 10 mol-% NEt$_3$ |
| 29 | **57** | NHMe$_2$ | PF$_6$ | 80 | 45 | 62:38 | |
| 30 | **57** | NHMe$_2$ | SbF$_6$ | 80 | 49 | 70:30 | |
| 31 | **57** | dbu | PF$_6$ | 80 | 61 | rac | |
| 32 | **57** | dbu | SbF$_6$ | 80 | 81 | rac | |
| 33 | **57** | NEt$_3$ | SbF$_6$ | 80 | 50 | 67:33 | |
| 34 | **57** | NEt$_3$ | SbF$_6$ | 80 | 40 | rac | 10 mol-% NEt$_3$ |
| 35 | **57** | NH$i$-Pr$_2$ | PF$_6$ | 80 | 55 | 68:32 | |
| 36 | **57** | NH$i$-Pr$_2$ | SbF$_6$ | 80 | 59 | 72:28 | |
| 37 | **57** | NH$i$-Pr$_2$ | SbF$_6$ | 80 | 61 | 72:28 | 10 mol-% SbF$_6$ |
| 38 | **57** | NH$_3$ | SbF$_6$ | 80 | 51 | 63:37 | |
| 39 | **57** | NH$_3$ | SbF$_6$ | 80 | 50 | 63:37 | 10 mol-% SbF$_6$ |
| 40 | **60** | - | - | 80 | 42 | rac | |
| 41 | **60** | NEt$_3$ | - | 40 | 19 | rac | |
| 42 | **60** | NEt$_3$ | - | 60 | 87 | rac | |
| 43 | **60** | NEt$_3$ | - | 80 | 78 | rac | |
| 44 | **60** | - | PF$_6$ | 60 | 15 | rac | |
| 45 | **60** | NEt$_3$ | SbF$_6$ | 80 | 35 | 58:42 | |
| 46 | **60** | NEt$_3$ | PF$_6$ | 80 | 23 | 56:44 | |
| 47 | **60** | NH$i$-Pr$_2$ | SbF$_6$ | 80 | 38 | 61:39 | |
| 48 | **58** | - | - | 80 | 4 | rac | |
| 49 | **58** | NEt$_3$ | - | 80 | 14 | rac | |
| 50 | **58** | - | PF$_6$ | 80 | 10 | rac | |
| 51 | **58** | NEt$_3$ | SbF$_6$ | 80 | 38 | 46:54 | |
| 52 | **58** | NH$_3$ | PF$_6$ | 80 | 40 | 44:56 | |
| 53 | **58** | NH$_3$ | SbF$_6$ | 80 | 59 | 40:60 | |
| 54 | **58** | NHMe$_2$ | SbF$_6$ | 80 | 81 | 42:58 | |
| 55 | **58** | NH$i$-Pr$_2$ | SbF$_6$ | 80 | 67 | 44:56 | |

## 5.3.6 Ru$^{II}$-Catalyzed Trifluoromethylation of β-Ketoesters

Ruthenium complex **57** (4.8 mg, 75 µmol, 10 mol%) and Et$_3$OPF$_6$ (3.8 mg, 155 µmol, 20 mol-%) were dissolved in CH$_2$Cl$_2$ (1 mL). After 90 min, the substrate (75 µmol) and Togni's reagent (0.1 M in CH$_2$Cl$_2$, 0.8 mL, 80 µmol, 1.05 eq) were added. A color change from dark red to orange was observed. As an internal standard, PhCF$_3$ was added (38 µL, 1.03 eq).

| entry | R$^{1/2}$ | R$^3$ | observation |
|---|---|---|---|
| 1 | Me / Me | 2,4,6-mesityl | no product formation |
| 2 | –(CH$_2$)$_3$– | Et | $^{19}$F-NMR signal at −70 ppm (small) |
| 3 | Ph / Me | Et | no product formation |

# Chapter 6

# Appendix

# 6.1 Crystallographic Data

## 6.1.1 ($S_p$)-1-Bromo-2-[(S)-1-phenylethyl]ferrocene (14)

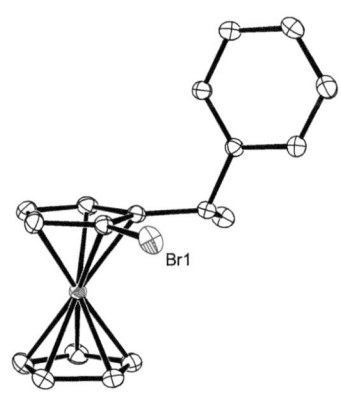

| | | | |
|---|---|---|---|
| identification code | ra004 | CCDC number | 909242 |
| cryst. method. | CH$_2$Cl$_2$ / EtOH | empirical formula | C$_{18}$H$_{17}$BrFe |
| shape | cube | moiety formula | C$_{18}$H$_{17}$BrFe |
| color | orange | $M_r$ | 369.08 |
| cryst size (mm) | 0.50 × 0.45 × 0.40 | T (K) | 100(2) |
| exp. time/frame (s) | 1 / 2 | solution method | direct |
| crystal system | orthorhombic | space group | $P2_12_12_1$ |
| $a$ (Å) | 7.5062(4) | $\alpha$ (°) | 90 |
| $b$ (Å) | 8.6541(5) | $\beta$ (°) | 90 |
| $c$ (Å) | 22.4258(13) | $\gamma$ (°) | 90 |
| $V$ (Å$^3$) | 1456.77(14) | $Z$ | 4 |
| $\rho_{calc}$ (g cm$^{-3}$) | 1.683 | $\mu$ (mm$^{-1}$) | 3.764 |
| $\theta_{min}$, $\theta_{max}$ (°) | 1.82, 35.64 | $F_{000}$ | 744 |
| limiting indices | $-12 \leq h \leq 12$ | data | 6712 |
| | $-14 \leq k \leq 13$ | restraints | 0 |
| | $-36 \leq l \leq 36$ | parameters | 182 |
| collected/unique refl. | 57509 / 6712 | $R_{int}$ | 0.0349 |
| $T_{max}$, $T_{min}$ | 0.314, 0.257 | $\Delta\rho_{max,min}$ (e Å$^{-3}$) | 0.92, −0.25 |
| final $R$ [$I > 2\sigma(I)$] | 0.022 | $S$ | 1.06 |
| final $R$ [all data] | 0.052 | Flack parameter | 0.017(4) |

## 6.1.2 ($R_p$)-1-Bromo-2-[($R$)-1-(1-naphthyl)ethyl]ferrocene (20)

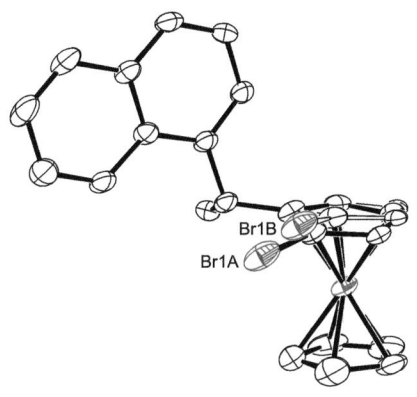

| | | | |
|---|---|---|---|
| identification code | ra087a | CCDC number | 909243 |
| cryst. method. | $CH_2Cl_2$ / EtOH | empirical formula | $C_{22}H_{19}BrFe$ |
| shape | prism | moiety formula | $C_{22}H_{19}BrFe$ |
| color | orange | $M_r$ | 419.13 |
| cryst size (mm) | $0.52 \times 0.50 \times 0.49$ | T (K) | 100(2) |
| exp. time/frame (s) | 0.5 | solution method | direct |
| crystal system | monoclinic | space group | $P2_1$ |
| $a$ (Å) | 9.455(3) | $\alpha$ (°) | 90 |
| $b$ (Å) | 7.698(3) | $\beta$ (°) | 101.336(7) |
| $c$ (Å) | 12.219(4) | $\gamma$ (°) | 90 |
| $V$ (Å$^3$) | 872.0(5) | $Z$ | 2 |
| $\rho_{calc}$ (g cm$^{-3}$) | 1.596 | $\mu$ (mm$^{-1}$) | 3.16 |
| $\theta_{min}, \theta_{max}$ (°) | 1.7, 28.7 | $F_{000}$ | 424 |
| limiting indices | $-12 \leq h \leq 12$ | data | 4363 |
| | $-10 \leq k \leq 10$ | restraints | 19 |
| | $-16 \leq l \leq 16$ | parameters | 255 |
| collected/unique refl. | 8602 / 4363 | $R_{int}$ | 0.065 |
| $T_{max}, T_{min}$ | 0.307, 0.291 | $\Delta\rho_{max,min}$ (e Å$^{-3}$) | 1.33, $-0.54$ |
| final $R$ [$I > 2\sigma(I)$] | 0.059 | $S$ | 0.98 |
| final $R$ [all data] | 0.15 | Flack parameter | 0.00(1) |

### 6.1.3 ($R_p$)-1-Bromo-2-[($R$)-1-(8-fluoronaphth-1-yl)ethyl]ferrocene (21)

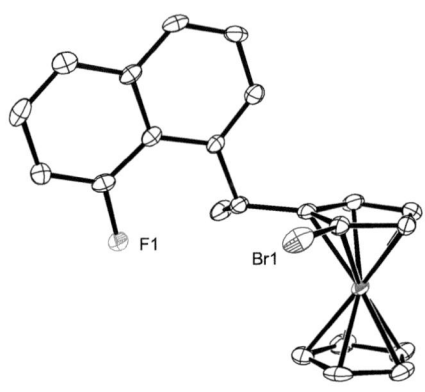

| | | | |
|---|---|---|---|
| identification code | ra105a | CCDC number | 909244 |
| cryst. method. | CH$_2$Cl$_2$ / EtOH | empirical formula | C$_{22}$H$_{18}$BrFFe |
| shape | prism | moiety formula | C$_{22}$H$_{18}$BrFFe |
| color | orange | $M_r$ | 437.12 |
| cryst size (mm) | 0.31 × 0.31 × 0.13 | T (K) | 100(2) |
| exp. time/frame (s) | 1 | solution method | direct |
| crystal system | monoclinic | space group | $P2_1$ |
| $a$ (Å) | 9.6503(11) | $\alpha$ (°) | 90 |
| $b$ (Å) | 7.3254(8) | $\beta$ (°) | 104.240(2) |
| $c$ (Å) | 12.9461(15) | $\gamma$ (°) | 90 |
| $V$ (Å$^3$) | 887.07(17) | $Z$ | 2 |
| $\rho_{calc}$ (g cm$^{-3}$) | 1.637 | $\mu$ (mm$^{-1}$) | 3.11 |
| $\theta_{min}$, $\theta_{max}$ (°) | 2.2, 28.3 | $F_{000}$ | 440 |
| limiting indices | $-12 \leq h \leq 12$ | data | 4334 |
| | $-9 \leq k \leq 9$ | restraints | 1 |
| | $-17 \leq l \leq 17$ | parameters | 227 |
| collected/unique refl. | 9237 / 4334 | $R_{int}$ | 0.034 |
| $T_{max}$, $T_{min}$ | 0.682, 0.450 | $\Delta\rho_{max,min}$ (e Å$^{-3}$) | 0.71, −0.40 |
| final $R$ [$I > 2\sigma(I)$] | 0.034 | $S$ | 0.98 |
| final $R$ [all data] | 0.071 | Flack parameter | 0.009(8) |

## 6.1.4 ($R_p$)-1-Bromo-2-[($R$)-1-(3,5-dimethylphenyl)ethyl]ferrocene (16)

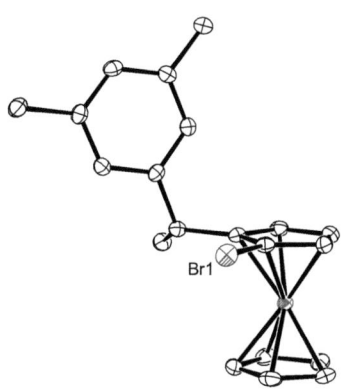

| | | | |
|---|---|---|---|
| identification code | es34 | CCDC number | 909245 |
| cryst. method. | $n$-hexane | empirical formula | $C_{20}H_{21}BrFe$ |
| shape | cube | moiety formula | $C_{20}H_{21}BrFe$ |
| color | orange | $M_r$ | 397.13 |
| cryst size (mm) | $0.58 \times 0.46 \times 0.42$ | T (K) | 100(2) |
| exp. time/frame (s) | 0.5 | solution method | direct |
| crystal system | orthorhombic | space group | $P2_12_12_1$ |
| $a$ (Å) | 8.3498(6) | $\alpha$ (°) | 90 |
| $b$ (Å) | 9.4035(7) | $\beta$ (°) | 90 |
| $c$ (Å) | 21.6218(15) | $\gamma$ (°) | 90 |
| $V$ (Å$^3$) | 1697.7 (2) | $Z$ | 4 |
| $\rho_{calc}$ (g cm$^{-3}$) | 1.554 | $\mu$ (mm$^{-1}$) | 3.24 |
| $\theta_{min}, \theta_{max}$ (°) | 1.9, 28.3 | $F_{000}$ | 808 |
| limiting indices | $-11 \leq h \leq 11$ | data | 4201 |
| | $-12 \leq k \leq 12$ | restraints | 0 |
| | $-28 \leq l \leq 28$ | parameters | 202 |
| collected/unique refl. | 17398 / 4201 | $R_{int}$ | 0.035 |
| $T_{max}, T_{min}$ | 0.344, 0.256 | $\Delta\rho_{max,min}$ (e Å$^{-3}$) | 0.49, $-0.25$ |
| final $R$ [$I > 2\sigma(I)$] | 0.024 | $S$ | 1.01 |
| final $R$ [all data] | 0.054 | Flack parameter | 0.004(7) |

### 6.1.5 $(R_p)$-1-Bromo-2-[$(R)$-1-(2,4,6-trimethylphenyl)ethyl]ferrocene (17)

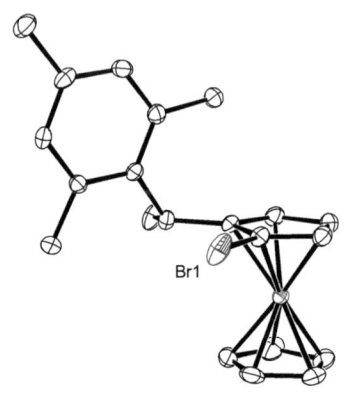

| | | | |
|---|---|---|---|
| identification code | ra091a | CCDC number | 909246 |
| cryst. method. | $CH_2Cl_2$ / $n$-hexane | empirical formula | $C_{21}H_{23}BrFe$ |
| shape | prism | moiety formula | $C_{21}H_{23}BrFe$ |
| color | orange | $M_r$ | 411.15 |
| cryst size (mm) | $0.47 \times 0.28 \times 0.21$ | T (K) | 100(2) |
| exp. time/frame (s) | 2 | solution method | direct |
| crystal system | monoclinic | space group | $P2_1$ |
| $a$ (Å) | 9.9716(9) | $\alpha$ (°) | 90 |
| $b$ (Å) | 12.1995(12) | $\beta$ (°) | 105.471(2) |
| $c$ (Å) | 15.3381(15) | $\gamma$ (°) | 90 |
| $V$ (Å$^3$) | 1798.2(3) | $Z$ | 4 |
| $\rho_{calc}$ (g cm$^{-3}$) | 1.519 | $\mu$ (mm$^{-1}$) | 3.06 |
| $\theta_{min}$, $\theta_{max}$ (°) | 1.4, 28.4 | $F_{000}$ | 840 |
| limiting indices | $-13 \leq h \leq 13$ | data | 8805 |
| | $-16 \leq k \leq 16$ | restraints | 1 |
| | $-20 \leq l \leq 20$ | parameters | 423 |
| collected/unique refl. | 18314 / 8805 | $R_{int}$ | 0.031 |
| $T_{max}$, $T_{min}$ | 0.566, 0.328 | $\Delta\rho_{max,min}$ (e Å$^{-3}$) | 0.59, $-0.58$ |
| final $R$ [$I > 2\sigma(I)$] | 0.030 | $S$ | 0.98 |
| final $R$ [all data] | 0.067 | Flack parameter | 0.014(6) |

## 6.1.6 ($R_p$)-1-Bromo-2-[($R$)-1-(3,5-bis(trifluoromethyl)phenyl)ethyl]ferrocene (19)

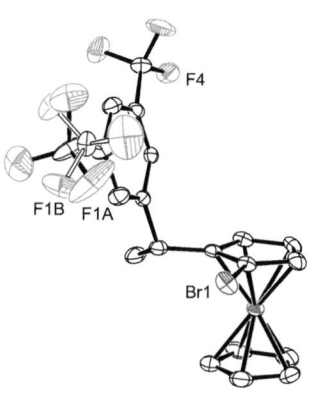

| | | | |
|---|---|---|---|
| identification code | es35 | CCDC number | 909247 |
| cryst. method. | $n$-hexane | empirical formula | $C_{20}H_{15}BrF_6Fe$ |
| shape | prism | moiety formula | $C_{20}H_{15}BrF_6Fe$ |
| color | orange | $M_r$ | 505.08 |
| cryst size (mm) | 0.41 × 0.35 × 0.26 | T (K) | 100(2) |
| exp. time/frame (s) | 0.5 | solution method | direct |
| crystal system | orthorhombic | space group | $P2_12_12_1$ |
| $a$ (Å) | 11.0459(8) | $\alpha$ (°) | 90 |
| $b$ (Å) | 11.6097(8) | $\beta$ (°) | 90 |
| $c$ (Å) | 14.7205(10) | $\gamma$ (°) | 90 |
| $V$ (Å$^3$) | 1887.8 (2) | $Z$ | 4 |
| $\rho_{calc}$ (g cm$^{-3}$) | 1.777 | $\mu$ (mm$^{-1}$) | 2.98 |
| $\theta_{min}, \theta_{max}$ (°) | 2.2, 25.4 | $F_{000}$ | 1000 |
| limiting indices | $-13 \leq h \leq 13$ | data | 3448 |
| | $-13 \leq k \leq 13$ | restraints | 18 |
| | $-17 \leq l \leq 17$ | parameters | 291 |
| collected/unique refl. | 15490 / 3448 | $R_{int}$ | 0.049 |
| $T_{max}, T_{min}$ | 0.510, 0.377 | $\Delta\rho_{max,min}$ (e Å$^{-3}$) | 0.30, −0.22 |
| final $R$ [$I > 2\sigma(I)$] | 0.028 | $S$ | 1.01 |
| final $R$ [all data] | 0.063 | Flack parameter | 0.025(9) |

## 6.1.7 ($S_p$)-1-Formyl-2-[(S)-1-phenylethyl]ferrocene (39)

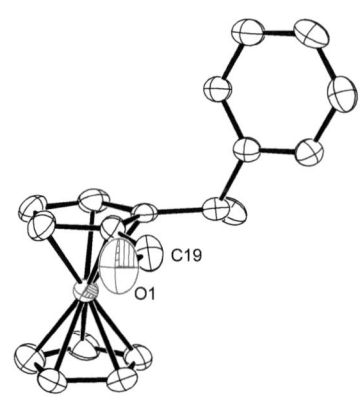

| | | | |
|---|---|---|---|
| identification code | ra016A | CCDC number | 909248 |
| cryst. method. | CH$_2$Cl$_2$/ $n$-hexane | empirical formula | C$_{19}$H$_{18}$FeO |
| shape | plate | moiety formula | C$_{19}$H$_{18}$FeO |
| color | orange | $M_r$ | 318.18 |
| cryst size (mm) | 0.19 × 0.17 × 0.09 | T (K) | 100(2) |
| exp. time/frame (s) | 4 | solution method | direct |
| crystal system | orthorhombic | space group | $P2_12_12_1$ |
| $a$ (Å) | 7.6281(10) | $\alpha$ (°) | 90 |
| $b$ (Å) | 8.5257(11) | $\beta$ (°) | 90 |
| $c$ (Å) | 22.957(3) | $\gamma$ (°) | 90 |
| $V$ (Å$^3$) | 1493.0(3) | $Z$ | 4 |
| $\rho_{calc}$ (g cm$^{-3}$) | 1.416 | $\mu$ (mm$^{-1}$) | 1.005 |
| $\theta_{min}$, $\theta_{max}$ (°) | 1.8, 28.4 | $F_{000}$ | 664 |
| limiting indices | $-10 \leq h \leq 10$ | data | 3738 |
| | $-11 \leq k \leq 11$ | restraints | 0 |
| | $-30 \leq l \leq 30$ | parameters | 190 |
| collected/unique refl. | 15675 / 3738 | $R_{int}$ | 0.0759 |
| $T_{max}$, $T_{min}$ | | $\Delta\rho_{max,min}$ (e Å$^{-3}$) | 0.711, −0.265 |
| final $R$ [$I > 2\sigma(I)$] | 0.0416 | $S$ | 0.814 |
| final $R$ [all data] | 0.661 | Flack parameter | −0.003(18) |

xx

## 6.1.8 $(S_p)$-Carboxyl-2-[$(S)$-1-phenylethyl]ferrocene (7)

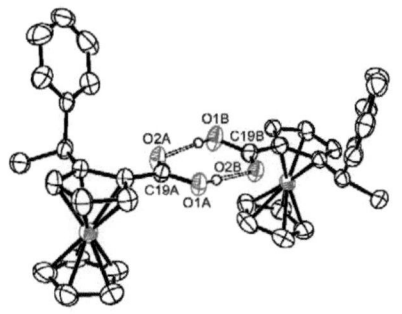

| | | | |
|---|---|---|---|
| identification code | RA_M_6_3 | CCDC number | 909249 |
| cryst. method. | EtOAc / $n$-hexane | empirical formula | $C_{38}H_{36}Fe_2O_4$ |
| shape | plate | moiety formula | $C_{19}H_{18}FeO_2$ |
| color | orange | $M_r$ | 668.37 |
| cryst size (mm) | $0.48 \times 0.20 \times 0.07$ | T (K) | 200(2) |
| exp. time/frame (s) | 6 | solution method | direct |
| crystal system | monoclinic | space group | $P2_1$ |
| $a$ (Å) | 7.7646(5) | $\alpha$ (°) | 90 |
| $b$ (Å) | 13.8431(9) | $\beta$ (°) | 100.770(2) |
| $c$ (Å) | 15.0584(9) | $\gamma$ (°) | 90 |
| $V$ (Å$^3$) | 1590.06(17) | $Z$ | 2 |
| $\rho_{calc}$ (g cm$^{-3}$) | 1.396 | $\mu$ (mm$^{-1}$) | 0.95 |
| $\theta_{min}$, $\theta_{max}$ (°) | 2.7, 28.3 | $F_{000}$ | 696 |
| limiting indices | $-10 \leq h \leq 10$ | data | 7828 |
| | $-18 \leq k \leq 18$ | restraints | 1 |
| | $-19 \leq l \leq 20$ | parameters | 405 |
| collected/unique refl. | 16630 / 7828 | $R_{int}$ | 0.038 |
| $T_{max}$, $T_{min}$ | 0.934, 0.659 | $\Delta\rho_{max,min}$ (e Å$^{-3}$) | 0.59, $-0.32$ |
| final $R$ [$I > 2\sigma(I)$] | 0.047 | $S$ | 1.01 |
| final $R$ [all data] | 0.097 | Flack parameter | $-0.004(13)$ |

## 6.1.9 $(S_p)$-N-(Benzyloxycarbonylamino)-2-[(S)-1-phenylethyl]ferrocene (8)

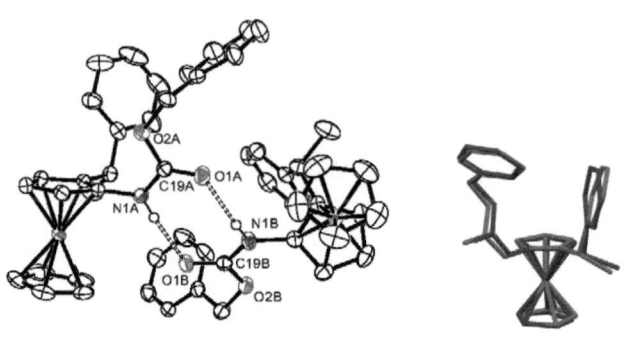

| | | | |
|---|---|---|---|
| identification code | RA_M_9_4 | CCDC number | 909250 |
| cryst. method. | $CH_2Cl_2$ / n-hexane | empirical formula | $C_{52}H_{50}Fe_2N_2O_4$ |
| shape | cube | moiety formula | $C_{26}H_{25}FeNO_2$ |
| color | orange | $M_r$ | 878.64 |
| cryst size (mm) | $0.51 \times 0.34 \times 0.19$ | T (K) | 200(2) |
| exp. time/frame (s) | 5 | solution method | direct |
| crystal system | orthorhombic | space group | $P2_12_12_1$ |
| a (Å) | 10.3546(8) | $\alpha$ (°) | 90 |
| b (Å) | 17.6796(14) | $\beta$ (°) | 90 |
| c (Å) | 23.5643(19) | $\gamma$ (°) | 90 |
| V (Å$^3$) | 4313.8(6) | Z | 4 |
| $\rho_{calc}$ (g cm$^{-3}$) | 1.353 | $\mu$ (mm$^{-1}$) | 0.72 |
| $\theta_{min}$, $\theta_{max}$ (°) | 1.4, 30.5 | $F_{000}$ | 1840 |
| limiting indices | $-14 \leq h \leq 14$ | data | 12622 |
| | $-24 \leq k \leq 24$ | restraints | 0 |
| | $-33 \leq l \leq 33$ | parameters | 541 |
| collected/unique refl. | 49409 / 12622 | $R_{int}$ | 0.051 |
| $T_{max}$, $T_{min}$ | 0.874, 0.708 | $\Delta\rho_{max,min}$ (e Å$^{-3}$) | 0.74, −0.22 |
| final R [$I > 2\sigma(I)$] | 0.046 | S | 1.06 |
| final R [all data] | 0.099 | Flack parameter | −0.001(9) |

## 6.1.10 ($S_p$)-2-[($S$)-1-phenylethyl]aminoferrocene (9)

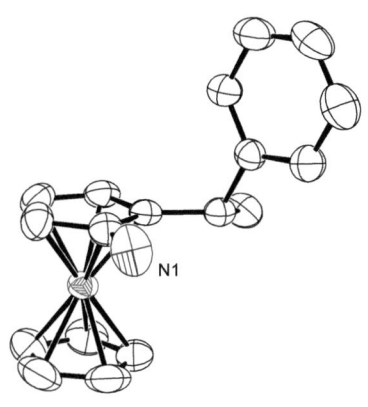

| | | | |
|---|---|---|---|
| identification code | RA_M_11_1 | CCDC number | 909251 |
| cryst. method. | $CH_2Cl_2$ / $n$-hexane | empirical formula | $C_{18}H_{19}FeN$ |
| shape | plate | moiety formula | $C_{18}H_{19}FeN$ |
| color | orange | $M_r$ | 305.19 |
| cryst size (mm) | 0.99 × 0.46 × 0.20 | T (K) | 293(2) |
| exp. time/frame (s) | 5 | solution method | direct |
| crystal system | orthorhombic | space group | $P2_12_12_1$ |
| $a$ (Å) | 7.5652(4) | $\alpha$ (°) | 90 |
| $b$ (Å) | 8.6389(5) | $\beta$ (°) | 90 |
| $c$ (Å) | 22.4178(13) | $\gamma$ (°) | 90 |
| $V$ (Å$^3$) | 1465.12(14) | $Z$ | 4 |
| $\rho_{calc}$ (g cm$^{-3}$) | 1.384 | $\mu$ (mm$^{-1}$) | 1.02 |
| $\theta_{min}, \theta_{max}$ (°) | 1.8, 28.3 | $F_{000}$ | 640 |
| limiting indices | $-9 \leq h \leq 10$ | data | 3624 |
| | $-11 \leq k \leq 11$ | restraints | 0 |
| | $-29 \leq l \leq 29$ | parameters | 181 |
| collected/unique refl. | 15187 / 3624 | $R_{int}$ | 0.026 |
| $T_{max}$, $T_{min}$ | 0.822, 0.433 | $\Delta\rho_{max,min}$ (e Å$^{-3}$) | 0.42, −0.18 |
| final $R$ [$I > 2\sigma(I)$] | 0.030 | $S$ | 1.09 |
| final $R$ [all data] | 0.071 | Flack parameter | 0.021(15) |

## 6.1.11 (*E*)-*N*-(((*Z*)-4-((*S*$_p$)-2-[(*S*)-1-phenylethyl]ferrocen-1-yl)amino)pent-3-en-2-yliden)-(*S*$_p$)-2-[(*S*)-1-phenyl-ethyl]aminoferrocene (29)

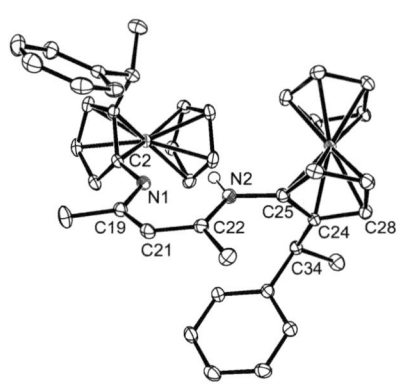

| | | | |
|---|---|---|---|
| identification code | ra039 | CCDC number | 909252 |
| cryst. method. | EtOAc / *n*-hexane | empirical formula | $C_{41}H_{42}Fe_2N_2$ |
| shape | prism | moiety formula | $C_{41}H_{42}Fe_2N_2$ |
| color | orange | $M_r$ | 674.47 |
| cryst size (mm) | 0.45 × 0.26 × 0.25 | T (K) | 100(2) |
| exp. time/frame (s) | 0.75 / 2 | solution method | direct |
| crystal system | monoclinic | space group | $P2_1$ |
| a (Å) | 8.1663(8) | α (°) | 90 |
| b (Å) | 21.561(2) | β (°) | 98.269(3) |
| c (Å) | 9.4028(9) | γ (°) | 90 |
| V (Å$^3$) | 1638.3(3) | Z | 2 |
| $\rho_{calc}$ (g cm$^{-3}$) | 1.367 | $\mu$ (mm$^{-1}$) | 0.92 |
| $\theta_{min}, \theta_{max}$ (°) | 1.9, 33.7 | $F_{000}$ | 708 |
| limiting indices | $-12 \leq h \leq 12$ | data | 13079 |
| | $-33 \leq k \leq 33$ | restraints | 1 |
| | $-14 \leq l \leq 14$ | parameters | 408 |
| collected/unique refl. | 61766 / 13079 | $R_{int}$ | 0.055 |
| $T_{max}, T_{min}$ | 0.804, 0.681 | $\Delta\rho_{max,min}$ (e Å$^{-3}$) | 1.26, −0.54 |
| final $R$ [$I > 2\sigma(I)$] | 0.038 | S | 1.05 |
| final $R$ [all data] | 0.088 | Flack parameter | 0.027(8) |

## 6.1.12 ($S_p$)-1-Diphenylphosphino-2-[($S$)-1-phenylethyl]ferrocene (40)

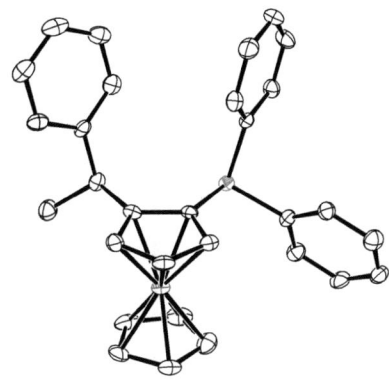

| | | | |
|---|---|---|---|
| identification code | ra062a | CCDC number | 909259 |
| cryst. method. | $CH_2Cl_2$ / $n$-hexane | empirical formula | $C_{30}H_{27}FeP$ |
| shape | cube | moiety formula | $C_{30}H_{27}FeP$ |
| color | orange | $M_r$ | 474.34 |
| cryst size (mm) | 0.19 × 0.12 × 0.11 | T (K) | 100(2) |
| exp. time/frame (s) | 4 | solution method | direct |
| crystal system | orthorhombic | space group | $P2_12_12_1$ |
| $a$ (Å) | 12.7876(13) | $\alpha$ (°) | 90 |
| $b$ (Å) | 13.2022(14) | $\beta$ (°) | 90 |
| $c$ (Å) | 13.6552(14) | $\gamma$ (°) | 90 |
| $V$ (Å$^3$) | 2305.3(4) | $Z$ | 4 |
| $\rho_{calc}$ (g cm$^{-3}$) | 1.337 | $\mu$ (mm$^{-1}$) | 0.74 |
| $\theta_{min}, \theta_{max}$ (°) | 2.2, 28.3 | $F_{000}$ | 992 |
| limiting indices | $-17 \leq h \leq 17$ | data | 5729 |
| | $-17 \leq k \leq 17$ | restraints | 0 |
| | $-18 \leq l \leq 18$ | parameters | 290 |
| collected/unique refl. | 23901 / 5729 | $R_{int}$ | 0.086 |
| $T_{max}, T_{min}$ | 0.926, 0.872 | $\Delta\rho_{max,min}$ (e Å$^{-3}$) | 0.58, −0.31 |
| final $R$ [$I > 2\sigma(I)$] | 0.055 | $S$ | 1.03 |
| final $R$ [all data] | 0.111 | Flack parameter | 0.039(19) |

## 6.1.13 $(R_p)$-1-Diphenylphosphino-2-[$(R)$-1-(1-naphthyl)ethyl]ferrocene (48)

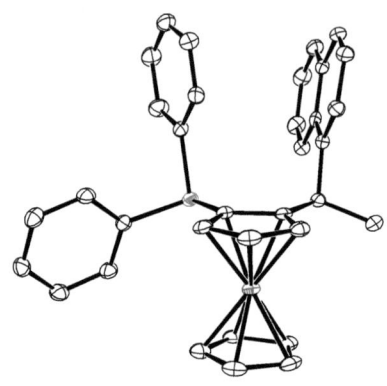

| | | | |
|---|---|---|---|
| identification code | ra088 | CCDC number | 909260 |
| cryst. method. | CH$_2$Cl$_2$ / $n$-hexane | empirical formula | C$_{34}$H$_{29}$FeP |
| shape | cube | moiety formula | C$_{34}$H$_{29}$FeP |
| color | orange | $M_r$ | 524.39 |
| cryst size (mm) | 0.70 × 0.64 × 0.49 | T (K) | 100(2) |
| exp. time/frame (s) | 0.5 | solution method | direct |
| crystal system | orthorhombic | space group | $P2_12_12_1$ |
| $a$ (Å) | 7.6957(18) | $\alpha$ (°) | 90 |
| $b$ (Å) | 13.322(3) | $\beta$ (°) | 90 |
| $c$ (Å) | 24.642(6) | $\gamma$ (°) | 90 |
| $V$ (Å$^3$) | 2526.4(10) | $Z$ | 4 |
| $\rho_{calc}$ (g cm$^{-3}$) | 1.379 | $\mu$ (mm$^{-1}$) | 0.68 |
| $\theta_{min}$, $\theta_{max}$ (°) | 1.7, 28.7 | $F_{000}$ | 1096 |
| limiting indices | $-10 \leq h \leq 10$ | data | 6412 |
| | $-17 \leq k \leq 17$ | restraints | 0 |
| | $-32 \leq l \leq 33$ | parameters | 326 |
| collected/unique refl. | 25608 / 6412 | $R_{int}$ | 0.064 |
| $T_{max}$, $T_{min}$ | 0.731, 0.645 | $\Delta\rho_{max,min}$ (e Å$^{-3}$) | 1.05, $-0.71$ |
| final $R$ [$I > 2\sigma(I)$] | 0.041 | $S$ | 1.03 |
| final $R$ [all data] | 0.099 | Flack parameter | 0.014(14) |

## 6.1.14 ($R_p$)-1-Diphenylphosphino-2-[($R$)-1-(3,5-dimethylphenyl)ethyl]ferrocene (44)

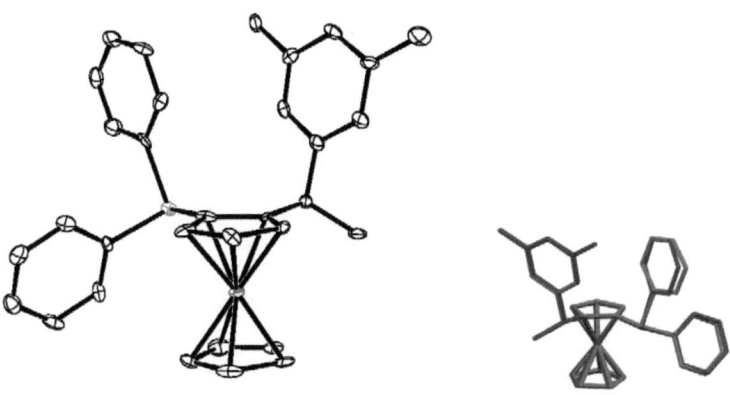

| | | | |
|---|---|---|---|
| identification code | es16 | CCDC number | 909261 |
| cryst. method. | $n$-hexane | empirical formula | $C_{32}H_{31}FeP$ |
| shape | prism | moiety formula | $C_{32}H_{31}FeP$ |
| color | orange | $M_r$ | 502.39 |
| cryst size (mm) | $0.27 \times 0.22 \times 0.21$ | T (K) | 100(2) |
| exp. time/frame (s) | 3 | solution method | direct |
| crystal system | monoclinic | space group | $P2_1$ |
| $a$ (Å) | 13.377(3) | $\alpha$ (°) | 90 |
| $b$ (Å) | 13.230(2) | $\beta$ (°) | 92.701(4) |
| $c$ (Å) | 13.922(3) | $\gamma$ (°) | 90 |
| $V$ (Å$^3$) | 2461.1(8) | $Z$ | 4 |
| $\rho_{calc}$ (g cm$^{-3}$) | 1.356 | $\mu$ (mm$^{-1}$) | 0.70 |
| $\theta_{min}, \theta_{max}$ (°) | 1.5, 27.1 | $F_{000}$ | 1056 |
| limiting indices | $-17 \leq h \leq 17$ | data | 10756 |
| | $-16 \leq k \leq 16$ | restraints | 73 |
| | $-17 \leq l \leq 17$ | parameters | 619 |
| collected/unique refl. | 22789 / 10756 | $R_{int}$ | 0.057 |
| $T_{max}, T_{min}$ | 0.868, 0.834 | $\Delta\rho_{max,min}$ (e Å$^{-3}$) | 4.2, −0.74 |
| final $R$ [$I > 2\sigma(I)$] | 0.098 | $S$ | 1.11 |
| final $R$ [all data] | 0.264 | Flack parameter | 0.09(3) |

## 6.1.15 ($R_p$)-1-Diphenylphosphino-2-[($R$)-1-(3,5-di-*tert*-butylphenyl)ethyl]ferrocene (46)

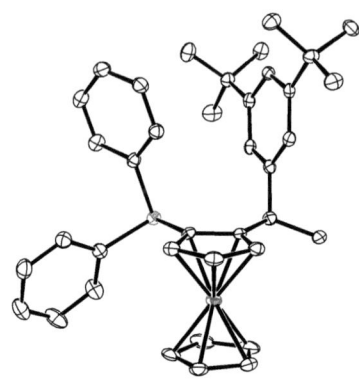

| | | | |
|---|---|---|---|
| identification code | ra094a | CCDC number | 909262 |
| cryst. method. | EtOH | empirical formula | $C_{38}H_{43}FeP$ |
| shape | needle | moiety formula | $C_{38}H_{43}FeP$ |
| color | orange | $M_r$ | 586.54 |
| cryst size (mm) | 0.55 × 0.28 × 0.11 | T (K) | 100(2) |
| exp. time/frame (s) | 2 | solution method | direct |
| crystal system | monoclinic | space group | $P2_1$ |
| a (Å) | 11.9031(14) | α (°) | 90 |
| b (Å) | 10.3079(12) | β (°) | 102.768(2) |
| c (Å) | 13.0922(15) | γ (°) | 90 |
| V (Å$^3$) | 1566.6(3) | Z | 2 |
| $\rho_{calc}$ (g cm$^{-3}$) | 1.243 | μ (mm$^{-1}$) | 0.557 |
| $\theta_{min}$, $\theta_{max}$ (°) | 1.6, 28.4 | $F_{000}$ | 624 |
| limiting indices | $-15 \leq h \leq 15$ | data | 7681 |
| | $-13 \leq k \leq 13$ | restraints | 1 |
| | $-17 \leq l \leq 17$ | parameters | 368 |
| collected/unique refl. | 16161 / 7681 | $R_{int}$ | 0.0421 |
| $T_{max}$, $T_{min}$ | 0.9423, 0.7492 | $\Delta\rho_{max,min}$ (e Å$^{-3}$) | 4.2, −0.74 |
| final R [$I > 2\sigma(I)$] | 0.042 | S | 0.995 |
| final R [all data] | 0.0928 | Flack parameter | 0.030(12) |

## 6.1.16 Chloro-(($S_p$)-1-diphenylphosphino-2-[($S$)-1-phenylethyl]ferrocenyl)gold(I) (51)

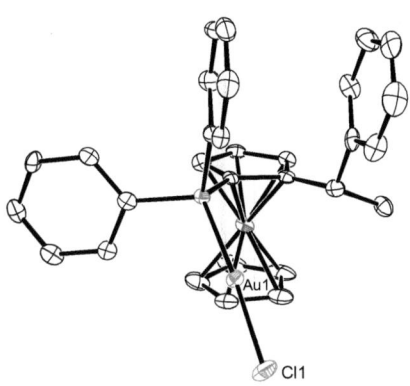

| | | | |
|---|---|---|---|
| identification code | ra063a | CCDC number | 909263 |
| cryst. method. | $CH_2Cl_2$ | empirical formula | $C_{31}H_{29}AuCl_3FeP$ |
| shape | needle | moiety formula | $C_{30}H_{27}AuClFeP$ $\cdot CH_2Cl_2$ |
| color | orange | $M_r$ | 791.68 |
| cryst size (mm) | $0.30 \times 0.13 \times 0.12$ | T (K) | 100(2) |
| exp. time/frame (s) | 3 | solution method | direct |
| crystal system | monoclinic | space group | $P2_1$ |
| a (Å) | 8.2603(9) | $\alpha$ (°) | 90 |
| b (Å) | 10.8911(12) | $\beta$ (°) | 100.931(2) |
| c (Å) | 16.2993(18) | $\gamma$ (°) | 90 |
| V (Å$^3$) | 1439.7(3) | Z | 2 |
| $\rho_{calc}$ (g cm$^{-3}$) | 1.826 | $\mu$ (mm$^{-1}$) | 5.95 |
| $\theta_{min}$, $\theta_{max}$ (°) | 2.3, 28.3 | $F_{000}$ | 772 |
| limiting indices | $-10 \leq h \leq 11$ | data | 6996 |
| | $-14 \leq k \leq 14$ | restraints | 1 |
| | $-21 \leq l \leq 21$ | parameters | 334 |
| collected/unique refl. | 14534 / 6996 | $R_{int}$ | 0.033 |
| $T_{max}$, $T_{min}$ | 0.541, 0.271 | $\Delta\rho_{max,min}$ (e Å$^{-3}$) | 2.66, $-1.35$ |
| final $R$ [$I > 2\sigma(I)$] | 0.034 | S | 1.02 |
| final $R$ [all data] | 0.072 | Flack parameter | 0.016(6) |

## 6.1.17 Chloro-(($R_p$)-1-diphenylphosphino-2-[($R$)-1-(1-naphthyl)ethyl]ferrocenyl)gold(I) (52)

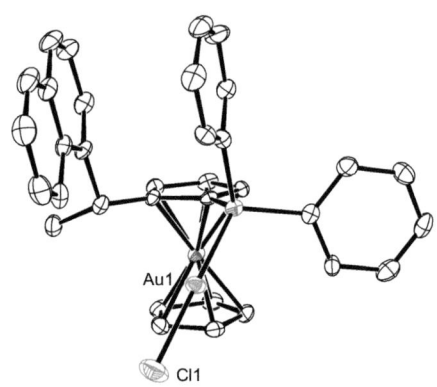

| | | | |
|---|---|---|---|
| identification code | ra089a | CCDC number | 909264 |
| cryst. method. | CH$_2$Cl$_2$ / EtOH | empirical formula | C$_{34}$H$_{29}$AuClFeP |
| shape | cube | moiety formula | C$_{34}$H$_{29}$AuClFeP |
| color | orange | $M_r$ | 756.81 |
| cryst size (mm) | 0.29 × 0.21 × 0.19 | T (K) | 100(2) |
| exp. time/frame (s) | 0.5 | solution method | direct |
| crystal system | orthorhombic | space group | $P2_12_12_1$ |
| $a$ (Å) | 8.2017(13) | $\alpha$ (°) | 90 |
| $b$ (Å) | 20.218(3) | $\beta$ (°) | 90 |
| $c$ (Å) | 20.514(3) | $\gamma$ (°) | 90 |
| $V$ (Å$^3$) | 3401.7(9) | $Z$ | 4 |
| $\rho_{calc}$ (g cm$^{-3}$) | 1.478 | $\mu$ (mm$^{-1}$) | 4.88 |
| $\theta_{min}$, $\theta_{max}$ (°) | 2.0, 28.4 | $F_{000}$ | 1480 |
| limiting indices | $-10 \leq h \leq 10$ | data | 8398 |
| | $-26 \leq k \leq 26$ | restraints | 0 |
| | $-27 \leq l \leq 27$ | parameters | 344 |
| collected/unique refl. | 34256 / 8398 | $R_{int}$ | 0.055 |
| $T_{max}$, $T_{min}$ | 0.456, 0.330 | $\Delta\rho_{max,min}$ (e Å$^{-3}$) | 1.71, −1.01 |
| final $R$ [$I > 2\sigma(I)$] | 0.031 | $S$ | 1.03 |
| final $R$ [all data] | 0.072 | Flack parameter | 0.029(5) |

## 6.1.18 Dichloro-(($R_p$)-1-κP-diphenylphosphino-2-[(S)-1-($\eta^6$-phenyl)ethyl]-ferrocenyl)ruthenium(II) (57)

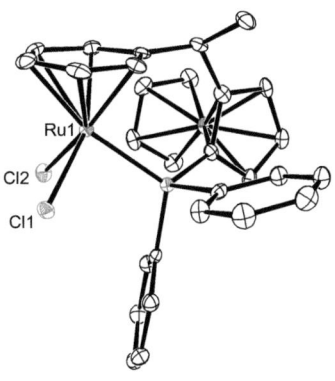

| | | | |
|---|---|---|---|
| identification code | ra072 | CCDC number | 909265 |
| cryst. method. | $CH_2Cl_2$ / EtOH | empirical formula | $C_{30}H_{27}Cl_2FePRu$ |
| shape | prism | moiety formula | $C_{30}H_{27}Cl_2FePRu$ |
| color | orange | $M_r$ | 646.31 |
| cryst size (mm) | 0.45 × 0.30 × 0.15 | T (K) | 100(2) |
| exp. time/frame (s) | 1 | solution method | direct |
| crystal system | orthorhombic | space group | $P2_12_12_1$ |
| a (Å) | 12.5699(16) | α (°) | 90 |
| b (Å) | 13.5885(17) | β (°) | 90 |
| c (Å) | 14.6810(18) | γ (°) | 90 |
| V (Å$^3$) | 2507.6(5) | Z | 4 |
| $\rho_{calc}$ (g cm$^{-3}$) | 1.712 | μ (mm$^{-1}$) | 1.48 |
| $\theta_{min}$, $\theta_{max}$ (°) | 2.0, 28.4 | $F_{000}$ | 1304 |
| limiting indices | $-16 \leq h \leq 16$ | data | 6259 |
| | $-18 \leq k \leq 18$ | restraints | 0 |
| | $-19 \leq l \leq 19$ | parameters | 317 |
| collected/unique refl. | 26189 / 6259 | $R_{int}$ | 0.054 |
| $T_{max}$, $T_{min}$ | 0.813, 0.556 | $\Delta\rho_{max,min}$ (e Å$^{-3}$) | 0.95, −0.47 |
| final R [$I > 2\sigma(I)$] | 0.031 | S | 1.02 |
| final R [all data] | 0.068 | Flack parameter | 0.020(16) |

## 6.1.19 Dichloro-(($R_p$)-1-$\kappa$P-diphenylphosphino-2-[($S$)-1-(3,5-dimethyl-($\eta^6$-phenyl))ethyl]-ferrocenyl)ruthenium(II) (60)

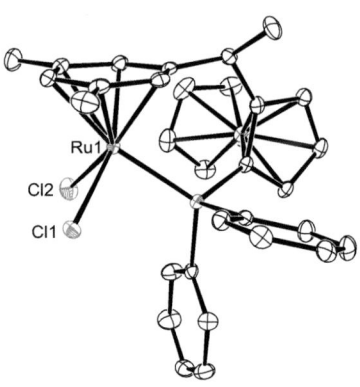

| | | | |
|---|---|---|---|
| identification code | raesx | CCDC number | 909266 |
| cryst. method. | CH$_2$Cl$_2$/ EtOH | empirical formula | C$_{32}$H$_{31}$Cl$_2$FePRu |
| shape | prism | moiety formula | C$_{32}$H$_{31}$Cl$_2$FePRu |
| color | red | $M_r$ | 674.36 |
| cryst size (mm) | 0.31 × 0.26 × 0.23 | T (K) | 100(2) |
| exp. time/frame (s) | 1 | solution method | direct |
| crystal system | orthorhombic | space group | $P2_12_12_1$ |
| $a$ (Å) | 9.4755(5) | $\alpha$ (°) | 90 |
| $b$ (Å) | 16.6580(8) | $\beta$ (°) | 90 |
| $c$ (Å) | 17.6248(9) | $\gamma$ (°) | 90 |
| $V$ (Å$^3$) | 2781.9(2) | $Z$ | 4 |
| $\rho_{calc}$ (g cm$^{-3}$) | 1.610 | $\mu$ (mm$^{-1}$) | 1.34 |
| $\theta_{min}$, $\theta_{max}$ (°) | 1.7, 29.6 | $F_{000}$ | 1368 |
| limiting indices | $-13 \leq h \leq 13$ | data | 8218 |
| | $-23 \leq k \leq 23$ | restraints | 0 |
| | $-24 \leq l \leq 24$ | parameters | 337 |
| collected/unique refl. | 42512/ 8218 | $R_{int}$ | 0.083 |
| $T_{max}$, $T_{min}$ | n.a. | $\Delta\rho_{max,min}$ (e Å$^{-3}$) | 0.81, $-0.67$ |
| final $R$ [$I > 2\sigma(I)$] | 0.034 | $S$ | 0.96 |
| final $R$ [all data] | 0.058 | Flack parameter | 0.007(14) |

## 6.1.20 Dichloro-(($S_p$)-1-$\kappa$P-diisopropylphosphino-2-[($R$)-1-($\eta^6$-phenyl)-ethyl]ferrocenyl)ruthenium(II) (58)

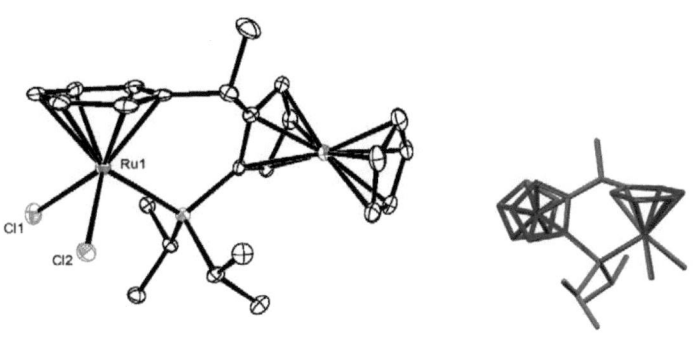

| | | | |
|---|---|---|---|
| identification code | ra136 | CCDC number | 909267 |
| cryst. method. | MeOH | empirical formula | $C_{24.10}H_{31.41}Cl_2FeO_{0.10}PRu$ |
| shape | plate | moiety formula | $C_{24}H_{31}Cl_2FePRu \cdot 0.1(CH_4O)$ |
| color | red | $M_r$ | 581.49 |
| cryst size (mm) | 0.47 × 0.38 × 0.10 | T (K) | 100(2) |
| exp. time/frame (s) | 1 | solution method | direct |
| crystal system | orthorhombic | space group | $P2_12_12_1$ |
| a (Å) | 9.8711(12) | $\alpha$ (°) | 90 |
| b (Å) | 12.9654(16) | $\beta$ (°) | 90 |
| c (Å) | 36.289(4) | $\gamma$ (°) | 90 |
| V (Å$^3$) | 4644.4(10) | Z | 8 |
| $\rho_{calc}$ (g cm$^{-3}$) | 1.663 | $\mu$ (mm$^{-1}$) | 1.58 |
| $\theta_{min}$, $\theta_{max}$ (°) | 1.7, 27.9 | $F_{000}$ | 2366 |
| limiting indices | $-12 \leq h \leq 12$ | data | 11075 |
| | $-16 \leq k \leq 17$ | restraints | 0 |
| | $-47 \leq l \leq 46$ | parameters | 544 |
| collected/unique refl. | 47096 / 11075 | $R_{int}$ | 0.054 |
| $T_{max}$, $T_{min}$ | 0.854, 0.521 | $\Delta\rho_{max,min}$ (e Å$^{-3}$) | 0.82, −0.42 |
| final $R$ [$I > 2\sigma(I)$] | 0.031 | S | 1.03 |
| final $R$ [all data] | 0.067 | Flack parameter | 0.015(13) |

## 6.1.21 Dichloro-(($R_p$)-1-$\kappa$P-diphenylphosphino-2-(($R$)-1-phenylethyl)ferrocenyl)[(1,2,3,4,5-$\eta$)-1,2,3,4,5-pentamethylcyclopentadienyl]iridium(III) (64)

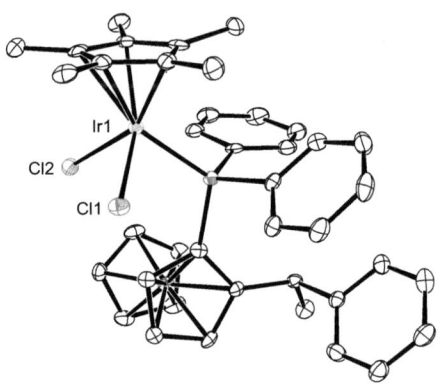

| | | | |
|---|---|---|---|
| identification code | ra085a | CCDC number | 909268 |
| cryst. method. | CH$_2$Cl$_2$ / $n$-hexane | empirical formula | C$_{41}$H$_{44}$Cl$_4$FeIrP |
| shape | cube | moiety formula | C$_{40}$H$_{42}$Cl$_2$FeIrP · CH$_2$Cl$_2$ |
| color | orange | $M_r$ | 957.58 |
| cryst size (mm) | 0.36 × 0.29 × 0.26 | T (K) | 100(2) |
| exp. time/frame (s) | 0.5 | solution method | Patterson |
| crystal system | orthorhombic | space group | $P2_12_12_1$ |
| $a$ (Å) | 11.6772(8) | $\alpha$ (°) | 90 |
| $b$ (Å) | 16.5561(12) | $\beta$ (°) | 90 |
| $c$ (Å) | 19.2174(13) | $\gamma$ (°) | 90 |
| $V$ (Å$^3$) | 3715.3(4) | $Z$ | 4 |
| $\rho_{calc}$ (g cm$^{-3}$) | 1.712 | $\mu$ (mm$^{-1}$) | 4.329 |
| $\theta_{min}$, $\theta_{max}$ (°) | 2.4, 26.2 | $F_{000}$ | 1904 |
| limiting indices | $-15 \leq h \leq 15$ | data | 9231 |
| | $-21 \leq k \leq 22$ | restraints | 0 |
| | $-25 \leq l \leq 24$ | parameters | 439 |
| collected/unique refl. | 38700 / 9231 | $R_{int}$ | 0.0627 |
| $T_{max}$, $T_{min}$ | 0.3956, 0.3033 | $\Delta\rho_{max,min}$ (e Å$^{-3}$) | 1.65, −0.61 |
| final $R$ [$I > 2\sigma(I)$] | 0.030 | $S$ | 0.92 |
| final $R$ [all data] | 0.060 | Flack parameter | −0.012(4) |

### 6.1.22 Dichloro-(($R_p$)-1-$\kappa$P-diphenylphosphino-2-(($R$)-1-phenylethyl)ferrocenyl)[(1,2,3,4,5-$\eta$)-1,2,3,4,5-pentamethyl-cyclopentadienyl]rhodium(III) (65)

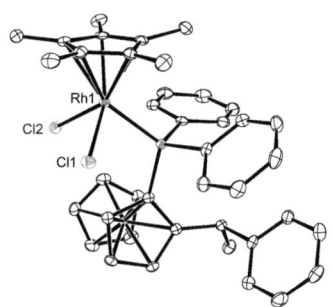

| | | | |
|---|---|---|---|
| identification code | ra108a | CCDC number | 909269 |
| cryst. method. | $CH_2Cl_2$ / $n$-hexane | empirical formula | $C_{41}H_{44}Cl_4FePRh$ |
| shape | prism | moiety formula | $C_{40}H_{42}Cl_2FePRh$ · $CH_2Cl_2$ |
| color | red | $M_r$ | 868.29 |
| cryst size (mm) | 0.40 × 0.32 × 0.26 | T (K) | 100(2) |
| exp. time/frame (s) | 1 | solution method | direct |
| crystal system | orthorhombic | space group | $P2_12_12_1$ |
| $a$ (Å) | 11.6671(12) | $\alpha$ (°) | 90 |
| $b$ (Å) | 16.5271(17) | $\beta$ (°) | 90 |
| $c$ (Å) | 19.188(2) | $\gamma$ (°) | 90 |
| $V$ (Å$^3$) | 3699.9(7) | $Z$ | 4 |
| $\rho_{calc}$ (g cm$^{-3}$) | 1.559 | $\mu$ (mm$^{-1}$) | 1.200 |
| $\theta_{min}$, $\theta_{max}$ (°) | 1.6, 28.3 | $F_{000}$ | 1776 |
| limiting indices | $-15 \leq h \leq 15$ | data | 9183 |
| | $-21 \leq k \leq 22$ | restraints | 0 |
| | $-25 \leq l \leq 25$ | parameters | 439 |
| collected/unique refl. | 38453 / 9183 | $R_{int}$ | 0.0424 |
| $T_{max}$, $T_{min}$ | 0.7400, 0.6486 | $\Delta\rho_{max,min}$ (e Å$^{-3}$) | 0.827, −0.324 |
| final $R$ [$I > 2\sigma(I)$] | 0.027 | $S$ | 1.021 |
| final $R$ [all data] | 0.0626 | Flack parameter | 0.000(12) |

## 6.1.23 $(S_p)$-1-Diphenylphosphino-2-[$(R)$-1-(1$H$-indol-1-yl)ethyl]ferrocene (66)

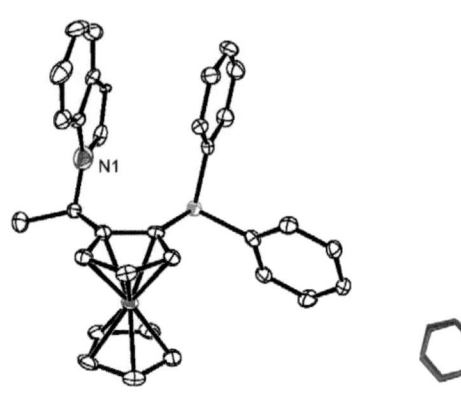

| | | | |
|---|---|---|---|
| identification code | ra138 | CCDC number | 909270 |
| cryst. method. | Et$_2$O/ EtOH | empirical formula | C$_{68}$H$_{62}$Fe$_2$N$_2$O$_2$P$_2$ |
| shape | plate | moiety formula | 2(C$_{32}$H$_{28}$FeNP) · 2(CH$_2$C$_6$O) |
| color | yellow | $M_r$ | 1112.84 |
| cryst size (mm) | 0.27 × 0.26 × 0.12 | T (K) | 100(2) |
| exp. time/frame (s) | 4 | solution method | direct |
| crystal system | monoclinic | space group | $P2_1$ |
| $a$ (Å) | 15.8344(15) | $\alpha$ (°) | 90 |
| $b$ (Å) | 9.3470(9) | $\beta$ (°) | 101.147(2) |
| $c$ (Å) | 19.4129(18) | $\gamma$ (°) | 90 |
| $V$ (Å$^3$) | 2819.0(5) | $Z$ | 2 |
| $\rho_{calc}$ (g cm$^{-3}$) | 1.311 | $\mu$ (mm$^{-1}$) | 0.619 |
| $\theta_{min}, \theta_{max}$ (°) | 1.1, 28.4 | $F_{000}$ | 1164 |
| limiting indices | $-21 \leq h \leq 21$ | data | 13984 |
| | $-12 \leq k \leq 12$ | restraints | 7 |
| | $-25 \leq l \leq 25$ | parameters | 707 |
| collected/unique refl. | 29564 / 13984 | $R_{int}$ | 0.0723 |
| $T_{max}, T_{min}$ | | $\Delta\rho_{max,min}$ (e Å$^{-3}$) | 0.617, −0.405 |
| final $R$ [$I > 2\sigma(I)$] | 0.0532 | $S$ | 0.866 |
| final $R$ [all data] | 0.1206 | Flack parameter | 0.002(14) |

# Bibliography

[1] Aardoom, R. Towards Ferrocenyl-Tethered Ligands for Asymmetric Transfer Hydrogenation. Master's Thesis, ETH Zurich, 2008.

[2] Kealy, T. J.; Pauson, P. L. *Nature* **1951**, *168*, 1039–1040.

[3] Laszlo, P.; Hoffmann, R. *Angew. Chem.* **2000**, *112*, 127–128.

[4] Togni, A.; Hayashi, T. Ferrocenes. VCH Verlagsgesellschaft mbH, Weinheim, 1995.

[5] Stepnicka, P. Ferrocenes: Ligands, Materials and Biomolecules. John Wiley & Sons, Ltd, Chichester, UK, 2008.

[6] Day, L.-X.; Hou, X.-L. Chiral Ferrocenes in Asymmetric Catalysis. Wiley-VCH Verlag GmbH & Co. KGaA, Weinheim, 2010.

[7] Werner, H. *Angew. Chem.* **2012**, *124*, 6156–6162.

[8] Woodward, R. B.; Rosenblum, M.; Whiting, M. C. *J. Am. Chem. Soc.* **1952**, *74*, 3458–3459.

[9] Wilkinson, G.; Rosenblum, M.; Whiting, M. C.; Woodward, R. B. *J. Am. Chem. Soc.* **1952**, *74*, 2125–2126.

[10] Fischer, E. O.; Pfab, W. *Z. Naturforsch. B: Chem. Sci.* **1952**, *7*, 377–379.

[11] Ruch, E.; Fischer, E. O. *Z. Naturforsch. B: Chem. Sci.* **1952**, *7*, 676–676.

[12] Dunitz, J. D.; Orgel, L. E. *Nature* **1953**, *171*, 121–122.

[13] Blaser, H.-U.; Brieden, W.; Pugin, B.; Spindler, F.; Studer, M.; Togni, A. *Top. Catal.* **2002**, *19*, 3–16.

[14] Zhou, Q.-L. Privileged Chiral Ligands and Catalysts. Wiley-VCH Verlag GmbH & Co. KGaA, Weinheim, 2011.

[15] Thomson, J. *Tetrahedron Lett.* **1959**, *1*, 26–27.

[16] Lednicer, D.; Hauser, C. R. *J. Org. Chem.* **1959**, *24*, 43–46.

[17] Cahn, R. S. *J. Chem. Educ.* **1964**, *41*, 116.

[18] Cahn, R. S.; Ingold, C.; Prelog, V. *Angew. Chem. Int. Ed. in Engl.* **1966**, *5*, 385–415.

[19] Prelog, V.; Helmchen, G. *Angew. Chem. Int. Ed. in Engl.* **1982**, *21*, 567–583.

[20] Schlögl, K. *Topics in Stereochemistry* **1967**, *1*, 39–91.

[21] Arimoto, F. S.; Haven, A. C. *J. Am. Chem. Soc.* **1955**, *77*, 6295–6297.

[22] Aratani, T.; Gonda, T.; Nozaki, H. *Tetrahedron Lett.* **1969**, *10*, 2265–2268.

[23] Marquarding, D.; Hoffmann, P.; Heitzer, H.; Ugi, I. *J. Am. Chem. Soc.* **1970**, *92*, 1969–1971.

[24] Goldberg, S. I.; Bailey, W. D. *J. Am. Chem. Soc.* **1971**, *93*, 1046–1047.

[25] Marquarding, D.; Klusacek, H.; Gokel, G.; Hoffmann, P.; Ugi, I. *Angew. Chem. Int. Ed. in Engl.* **1970**, *9*, 371–372.

[26] Marquarding, D.; Klusacek, H.; Gokel, G.; Hoffmann, P.; Ugi, I. *J. Am. Chem. Soc.* **1970**, *92*, 5389–5393.

[27] Gokel, G. W.; Ugi, I. K. *J. Chem. Educ.* **1972**, *49*, 294–296.

[28] Rebière, F.; Riant, O.; Ricard, L.; Kagan, H. B. *Angew. Chem.* **1993**, *105*, 644–646.

[29] Riant, O.; Samuel, O.; Kagan, H. B. *J. Am. Chem. Soc.* **1993**, *115*, 5835–5836.

[30] Sammakia, T.; Latham, H. A.; Schaad, D. R. *J. Org. Chem.* **1995**, *60*, 10–11.

[31] Gokel, G.; Hoffmann, P.; Klusacek, H.; Marquarding, D.; Ruch, E.; Ugi, I. *Angew. Chem. Int. Ed. in Engl.* **1970**, *9*, 64–65.

[32] Gleiter, R.; Bleiholder, C.; Rominger, F. *Organometallics* **2007**, *26*, 4850–4859.

[33] Takahashi, T.; Konno, T.; Ogata, K.; Fukuzawa, S.-i. *J. Org. Chem.* **2012**, *77*, 6638–6642.

[34] Togni, A.; Breutel, C.; Schnyder, A.; Spindler, F.; Landert, H.; Tijani, A. *J. Am. Chem. Soc.* **1994**, *116*, 4062–4066.

[35] Meerwein, H.; Schmidt, R. *Ann.* **1925**, *444*, 221–238.

[36] Ponndorf, W. *Angew. Chem.* **1926**, *39*, 138–143.

[37] Verley, A. *Bull. Soc. Chim. Fr.* **1937**, *37*, 871–874.

[38] Oppenauer, R. V. *Rec. Trav. Chim. Pays-Bas* **1937**, *56*, 137–144.

[39] Adkins, H.; Elofson, R. M.; Rossow, A. G.; Robinson, C. C. *J. Am. Chem. Soc.* **1949**, *71*, 3622–3629.

[40] Grigg, R.; Mitchell, T. R.; Sutthivaiyakit, S. *Tetrahedron* **1981**, *37*, 4313–4319.

[41] Murahashi, S.; Naota, T.; Ito, K.; Maeda, Y.; Taki, H. *J. Org. Chem.* **1987**, *52*, 4319–4327.

[42] Maytum, H. C.; Tavassoli, B.; Williams, J. M. J. *Org. Lett.* **2007**, *9*, 4387–4389.

[43] Zweifel, T.; Naubron, J.-V.; Büttner, T.; Ott, T.; Grützmacher, H. *Angew. Chem. Int. Ed.* **2008**, *47*, 3245–3249.

[44] Lundberg, H.; Adolfsson, H. *Tetrahedron Lett.* **2011**, *52*, 2754–2758.

[45] Yang, J. W.; List, B. *Org. Lett.* **2006**, *8*, 5653–5655.

[46] Zassinovich, G.; Mestroni, G.; Gladiali, S. *Chem. Rev.* **1992**, *92*, 1051–1069.

[47] Gladiali, S.; Alberico, E. *Chem. Soc. Rev.* **2006**, *35*, 226–236.

[48] Samec, J. S. M.; Backvall, J.-E.; Andersson, P. G.; Brandt, P. *Chem. Soc. Rev.* **2006**, *35*, 237–248.

[49] Ikariya, T.; Blacker, A. J. *Acc. Chem. Res.* **2007**, *40*, 1300–1308.

[50] Wang, C.; Wu, X.; Xiao, J. *Chem. Asian J.* **2008**, *3*, 1750–1770.

[51] Doering, W. v. E.; Young, R. W. *J. Am. Chem. Soc.* **1950**, *72*, 631–631.

[52] Hashiguchi, S.; Fujii, A.; Takehara, J.; Ikariya, T.; Noyori, R. *J. Am. Chem. Soc.* **1995**, *117*, 7562–7563.

[53] Müller, D.; Umbricht, G.; Weber, B.; Pfaltz, A. *Helv. Chim. Acta* **1991**, *74*, 232–240.

[54] Evans, D. A.; Nelson, S. G.; Gagne, M. R.; Muci, A. R. *J. Am. Chem. Soc.* **1993**, *115*, 9800–9801.

[55] Genêt, J.-P.; Ratovelomanana-Vidal, V.; Pinel, C. *Synlett* **1993**, *1993*, 478–480.

[56] Gamez, P.; Fache, F.; Lemaire, M. *Tetrahedron: Asymmetry* **1995**, *6*, 705–718.

[57] Yamakawa, M.; Ito, H.; Noyori, R. *J. Am. Chem. Soc.* **2000**, *122*, 1466–1478.

[58] Gao, J.-X.; Ikariya, T.; Noyori, R. *Organometallics* **1996**, *15*, 1087–1089.

[59] Püntener, K.; Schwink, L.; Knochel, P. *Tetrahedron Lett.* **1996**, *37*, 8165–8168.

[60] Jiang, Y.; Jiang, Q.; Zhang, X. *J. Am. Chem. Soc.* **1998**, *120*, 3817–3818.

[61] Alonso, D. A.; Nordin, S. J. M.; Roth, P.; Tarnai, T.; Andersson, P. G.; Thommen, M.; Pittelkow, U. *J. Org. Chem.* **2000**, *65*, 3116–3122.

[62] Hannedouche, J.; Clarkson, G. J.; Wills, M. *J. Am. Chem. Soc.* **2004**, *126*, 986–987.

[63] Hayes, A. M.; Morris, D. J.; Clarkson, G. J.; Wills, M. *J. Am. Chem. Soc.* **2005**, *127*, 7318–7319.

[64] Nesmeyanov, A. N.; Perevalova, E. G.; Golovnya, R. V.; Shilovtseva, L. S. *Dokl. Akad. Nauk. SSSR* **1955**, *102*, 535–538.

[65] van Leusen, D.; Hessen, B. *Organometallics* **2001**, *20*, 224–226.

[66] Nesmeyanov, A. N.; Ssasonova, V. A.; Gerasimenko, V.; Medvedeva, V. G. *Izv. Akad. Nauk. SSSR (Engl. Trans.)* **1962**, *4*, 1980–1984.

[67] Shafir, A.; Power, M. P.; Whitener, G. D.; Arnold, J. *Organometallics* **2000**, *19*, 3978–3982.

[68] Salter, R.; Pickett, T. E.; Richards, C. J. *Tetrahedron: Asymmetry* **1998**, *9*, 4239–4247.

[69] Riant, O.; Samuel, O.; Flessner, T.; Taudien, S.; Kagan, H. B. *J. Org. Chem.* **1997**, *62*, 6733–6745.

[70] Nilewski, C.; Neumann, M.; Tebben, L.; Fröhlich, R.; Kehr, G.; Erker, G. *Synthesis* **2006**, *13*, 2191–2200.

[71] McGinnis, D. M.; Deplazes, S. F.; Barybin, M. V. *J. Organomet. Chem.* **2011**, *696*, 3939–3944.

[72] Metallinos, C.; John, J.; Zaifman, J.; Emberson, K. *Adv. Synth. Catal.* **2012**, *354*, 602–606.

[73] Bertogg, A.; Camponovo, F.; Togni, A. *Eur. J. Inorg. Chem.* **2005**, *2005*, 347–356.

[74] Bertogg, A.; Togni, A. *Organometallics* **2006**, *25*, 622–630.

[75] Lee, S.-M.; Kowallick, R.; Marcaccio, M.; McCleverty, J. A.; Ward, M. D. *J. Chem. Soc., Dalton Trans.* **1998**, 3443–3450.

[76] Plenio, H.; Aberle, C. *Chem. Eur. J.* **2001**, *7*, 4438–4446.

[77] Bertogg, A. Planar-chirale Aminoferrocene: Bausteine zur Synthese neuartiger Amidinato- und Carbenliganden für die asymmetrische Katalyse. Diss. ETH No. 16642, ETH Zurich, 2006.

[78] Gibson, V. C.; Gregson, C. K.; Halliwell, C. M.; Long, N. J.; Oxford, P. J.; White, A. J.; Williams, D. J. *J. Organomet. Chem.* **2005**, *690*, 6271–6283.

[79] Metallinos, C.; Belle, L. V. *J. Organomet. Chem.* **2011**, *696*, 141–149.

[80] Tappe, K.; Knochel, P. *Tetrahedron: Asymmetry* **2004**, *15*, 91–102.

[81] Han, J.; Tokunaga, N.; Hayashi, T. *Helv. Chim. Acta* **2002**, *85*, 3848–3854.

[82] Barbaro, P.; Bianchini, C.; Giambastiani, G.; Togni, A. *Tetrahedron Lett.* **2003**, *44*, 8279–8283.

[83] Camponovo, F. Chiral Ferrocenyl Amidines as Modular Ligands for Applications in Asymmetric Catalysis. Diss. ETH No. 18199, ETH Zurich, 2009.

[84] Heinze, K.; Schlenker, M. *Eur. J. Inorg. Chem.* **2004**, *2004*, 2974–2988.

[85] Guram, A. S.; Rennels, R. A.; Buchwald, S. L. *Angew. Chem. Int. Ed. in Engl.* **1995**, *34*, 1348–1350.

[86] Louie, J.; Hartwig, J. F. *Tetrahedron Lett.* **1995**, *36*, 3609 – 3612.

[87] Scheibe, G. *Ber.* **1923**, *56*, 137–148.

[88] Schwarzenbach, G.; Lutz, K. *Helv. Chim. Acta* **1940**, *23*, 1139–1146.

[89] McGeachin, S. G. *Can. J. Chem.* **1968**, *46*, 1903–1912.

[90] Bradley, W.; Wright, I. *J. Chem. Soc.* **1956**, 640–648.

[91] Dorman, L. C. *Tetrahedron Lett.* **1966**, *7*, 459–464.

[92] Fritschi, H.; Leutenegger, U.; Pfaltz, A. *Angew. Chem. Int. Ed. in Engl.* **1986**, *25*, 1005–1006.

[93] Lowenthal, R. E.; Abiko, A.; Masamune, S. *Tetrahedron Lett.* **1990**, *31*, 6005–6008.

[94] Evans, D. A.; Woerpel, K. A.; Hinman, M. M.; Faul, M. M. *J. Am. Chem. Soc.* **1991**, *113*, 726–728.

[95] Corey, E. J.; Imai, N.; Zhang, H. Y. *J. Am. Chem. Soc.* **1991**, *113*, 728–729.

[96] Li, Z.; Conser, K. R.; Jacobsen, E. N. *J. Am. Chem. Soc.* **1993**, *115*, 5326–5327.

[97] Johnson, L. K.; Killian, C. M.; Brookhart, M. *J. Am. Chem. Soc.* **1995**, *117*, 6414–6415.

[98] Harder, S.; Buch, F. *Z. Naturforsch. B: Chem. Sci.* **2008**, *63*, 169–177.

[99] Buch, F. Erste Entwicklungen in der calciumbasierten Katalyse. Diss. Universität Duisburg, Essen, 2008.

[100] Oguadinma, P. O.; Schaper, F. *Organometallics* **2009**, *28*, 4089–4097.

[101] Bourget-Merle, L.; Lappert, M. F.; Severn, J. R. *Chem. Rev.* **2002**, *102*, 3031–3066.

[102] Tsai, Y.-C. *Coord. Chem. Rev.* **2012**, *256*, 722–758.

[103] Crimmin, M. R.; Casely, I. J.; Hill, M. S. *J. Am. Chem. Soc.* **2005**, *127*, 2042–2043.

[104] Bernskoetter, W. H.; Lobkovsky, E.; Chirik, P. J. *Chem. Commun.* **2004**, 764–765.

[105] Masuda, J. D.; Stephan, D. W. *Can. J. Chem.* **2005**, *83*, 324–327.

[106] Monillas, W. H.; Yap, G. P.; Theopold, K. H. *Inorg. Chim. Acta* **2011**, *369*, 103–119.

[107] Guan, B.; Xing, D.; Cai, G.; Wan, X.; Yu, N.; Fang, Z.; Yang, L.; Shi, Z. *J. Am. Chem. Soc.* **2005**, *127*, 18004–18005.

[108] Phillips, A. D.; Laurenczy, G.; Scopelliti, R.; Dyson, P. J. *Organometallics* **2007**, *26*, 1120–1122.

[109] Moreno, A.; Pregosin, P. S.; Laurenczy, G.; Phillips, A. D.; Dyson, P. J. *Organometallics* **2009**, *28*, 6432–6441.

[110] Phillips, A. D.; Zava, O.; Scopelitti, R.; Nazarov, A. A.; Dyson, P. J. *Organometallics* **2010**, *29*, 417–427.

[111] Drouin, F.; Oguadinma, P. O.; Whitehorne, T. J. J.; Prud'homme, R. E.; Schaper, F. *Organometallics* **2010**, *29*, 2139–2147.

[112] Stanlake, L. J. E.; Stephan, D. W. *Dalton Trans.* **2011**, *40*, 5836–5840.

[113] Aso, Y.; Yamashita, H.; Otsubo, T.; Ogura, F. *J. Org. Chem.* **1989**, *54*, 5627–5629.

[114] Hadzovic, A.; Song, D. *Inorg. Chem.* **2008**, *47*, 12010–12017.

[115] Yokota, S.; Tachi, Y.; Itoh, S. *Inorganic Chemistry* **2002**, *41*, 1342–1344.

[116] Stender, M.; Wright, R. J.; Eichler, B. E.; Prust, J.; Olmstead, M. M.; Roesky, H. W.; Power, P. P. *J. Chem. Soc., Dalton Trans.* **2001**, 3465–3469.

[117] Bondi, A. *J. Phys. Chem.* **1964**, *68*, 441–451.

[118] Rowland, R. S.; Taylor, R. *J. Phys. Chem.* **1996**, *100*, 7384–7391.

[119] Flückiger, M. Iron(II) in Asymmetric Hydrosilylation. Diss. ETH No. 19862, ETH Zurich, 2011.

[120] Wöhler, F. *Ann. Phys.* **1828**, *88*, 253–256.

[121] Volz, N.; Clayden, J. *Angew. Chem. Int. Ed.* **2011**, *50*, 12148–12155.

[122] Ishikawa, T. Superbases for Organic Synthesis. John Wiley & Sons, Ltd, Chichester, UK, 2009.

[123] Curran, D. P.; Kuo, L. H. *J. Org. Chem.* **1994**, *59*, 3259–3261.

[124] Wang, J.; Li, H.; Zu, L.; Jiang, W.; Xie, H.; Duan, W.; Wang, W. *J. Am. Chem. Soc.* **2006**, *128*, 12652–12653.

[125] Biddle, M. M.; Lin, M.; Scheidt, K. A. *J. Am. Chem. Soc.* **2007**, *129*, 3830–3831.

[126] Hoashi, Y.; Okino, T.; Takemoto, Y. *Angew. Chem.* **2005**, *117*, 4100–4103.

[127] Inokuma, T.; Hoashi, Y.; Takemoto, Y. *J. Am. Chem. Soc.* **2006**, *128*, 9413–9419.

[128] Okino, T.; Hoashi, Y.; Takemoto, Y. *J. Am. Chem. Soc.* **2003**, *125*, 12672–12673.

[129] Okino, T.; Hoashi, Y.; Furukawa, T.; Xu, X.; Takemoto, Y. *J. Am. Chem. Soc.* **2005**, *127*, 119–125.

[130] Huang, H.; Jacobsen, E. N. *J. Am. Chem. Soc.* **2006**, *128*, 7170–7171.

[131] Lalonde, M. P.; Chen, Y.; Jacobsen, E. N. *Angew. Chem.* **2006**, *118*, 6514–6518.

[132] Tsogoeva, S. B.; Wei, S. *Chem. Commun.* **2006**, 1451–1453.

[133] Li, H.; Wang, J.; Zu, L.; Wang, W. *Tetrahedron Lett.* **2006**, *47*, 2585–2589.

[134] Hynes, P. S.; Stranges, D.; Stupple, P. A.; Guarna, A.; Dixon, D. J. *Org. Lett.* **2007**, *9*, 2107–2110.

[135] Lubkoll, J.; Wennemers, H. *Angew. Chem.* **2007**, *119*, 6965–6968.

[136] Wei, S.; Yalalov, D. A.; Tsogoeva, S. B.; Schmatz, S. *Cat. Today* **2007**, *121*, 151–157.

[137] Andrés, J.; Manzano, R.; Pedrosa, R. *Chem. Eur. J.* **2008**, *14*, 5116–5119.

[138] Tan, K.; Jacobsen, E. *Angew. Chem. Int. Ed.* **2007**, *46*, 1315–1317.

[139] Sigrist, L. Synthesis of Novel Ferrocenylphosphines and Their Application in Gold(I)-catalyzed Asymmetric Intramolecular Hydroamination. Research Project I, ETH Zurich, 2010.

[140] Schneider, E. Ferrocenyl-Tethered Ruthenium-Arene Compexes and Their Application in Asymmetric Catalysis. Research Project I, ETH Zurich, 2011.

[141] Atkinson, C. J.; Long, N. J. Ferrocenes. John Wiley & Sons, Ltd, Chichester, UK, 2008.

[142] Dai, L.-X.; Hou, X.-L. Chiral Ferrocenes in Asymmetric Catalysis. Wiley-VCH Verlag GmbH & Co. KGaA, Weinheim, 2010.

[143] Pereira, S. I.; Adrio, J.; Silva, A. M. S.; Carretero, J. C. *J. Org. Chem.* **2005**, *70*, 10175–10177.

[144] Thalji, R. K.; Ahrendt, K. A.; Bergman, R. G.; Ellman, J. A. *J. Org. Chem.* **2005**, *70*, 6775–6781.

[145] Juge, S.; Stephan, M.; Laffitte, J. A.; Genet, J. *Tetrahedron Lett.* **1990**, *31*, 6357–6360.

[146] Colby, E. A.; Jamison, T. F. *J. Org. Chem.* **2003**, *68*, 156–166.

[147] Nettekoven, U.; Kamer, P. C. J.; van Leeuwen, P. W. N. M.; Widhalm, M.; Spek, A. L.; Lutz, M. *J. Org. Chem.* **1999**, *64*, 3996–4004.

[148] Maienza, F.; Wörle, M.; Steffanut, P.; Mezzetti, A.; Spindler, F. *Organometallics* **1999**, *18*, 1041–1049.

[149] Pedersen, H. L.; Johannsen, M. *J. Org. Chem.* **2002**, *67*, 7982–7994.

[150] Miller, K. M.; Jamison, T. F. *Org. Lett.* **2005**, *7*, 3077–3080.

[151] Jensen, J. F.; Johannsen, M. *Org. Lett.* **2003**, *5*, 3025–3028.

[152] Patel, S. J.; Jamison, T. F. *Angew. Chem.* **2004**, *116*, 4031–4034.

[153] Pearson, R. G. *J. Am. Chem. Soc.* **1963**, *85*, 3533–3539.

[154] Pearson, R. G. Chemical Hardness. Wiley-VCH Verlag GmbH & Co. KGaA, Weinheim, 2005.

[155] Bürgler, J. Bidentate and Tridentate P-Stereogenic Ferrocenyl Phosphines. Diss. ETH No. 119513, ETH Zurich, 2011.

[156] Yamamoto, H.; Oshima, K. Main Group Metals in Organic Synthesis. Wiley-VCH Verlag GmbH & Co. KGaA, Weinheim, 2005.

[157] Hunter, C. A.; Sanders, J. K. M. *J. Am. Chem. Soc.* **1990**, *112*, 5525–5534.

[158] Meyer, E. A.; Castellano, R. K.; Diederich, F. *Angew. Chem. Int. Ed.* **2003**, *42*, 1210–1250.

[159] Schwemberger, W.; Gordon, W. *Chem. Zentralb.* **1935**, *106*, 514.

[160] Ito, Y.; Sawamura, M.; Hayashi, T. *J. Am. Chem. Soc.* **1986**, *108*, 6405–6406.

[161] Hashmi, A. S. K.; Toste, F. D. Modern Gold Catalyzed Synthesis. Wiley-VCH Verlag GmbH & Co. KGaA, Weinheim, 2012.

[162] Laguna, A. Modern Supramolecular Gold Chemistry. Wiley-VCH Verlag GmbH & Co. KGaA, Weinheim, 2008.

[163] Hashmi, A. S. K. *Gold Bull.* **2004**, *37*, 51–65.

[164] Hashmi, A. S. K. *Angew. Chem. Int. Ed.* **2005**, *44*, 6990–6993.

[165] Li, Z.; Brouwer, C.; He, C. *Chem. Rev.* **2008**, *108*, 3239–3265.

[166] Widenhoefer, R. *Chem. Eur. J.* **2008**, *14*, 5382–5391.

[167] Sengupta, S.; Shi, X. *ChemCatChem* **2010**, *2*, 609–619.

[168] Pradal, A.; Toullec, Y. Patrick; Michelet, V. *Synthesis* **2011**, *2011*, 1501–1514.

[169] Gockel, B.; Krause, N. *Org. Lett.* **2006**, *8*, 4485–4488.

[170] Patil, N. T.; Lutete, L. M.; Nishina, N.; Yamamoto, Y. *Tetrahedron Lett.* **2006**, *47*, 4749–4751.

[171] Sherry, B. D.; Toste, F. D. *J. Am. Chem. Soc.* **2004**, *126*, 15978–15979.

[172] Shi, X.; Gorin, D. J.; Toste, F. D. *J. Am. Chem. Soc.* **2005**, *127*, 5802–5803.

[173] Bats, J. W.; Hamzic, M.; Hashmi, A. S. K. *Acta Cryst., Sect. E* **2007**, *63*, m2344.

[174] Gimeno, M. C.; Laguna, A.; Sarroca, C.; Jones, P. G. *Inorg. Chem.* **1993**, *32*, 5926–5932.

[175] Garcia-Seijo, M. I.; Sevillano, P.; Gould, R. O.; Fernandez-Anca, D.; Garcia-Fernandez, M. E. *Inorg. Chim. Acta* **2003**, *353*, 206–216.

[176] Liu, C.; Widenhoefer, R. A. *Org. Lett.* **2007**, *9*, 1935–1938.

[177] Teller, H.; Flügge, S.; Goddard, R.; Fürstner, A. *Angew. Chem. Int. Ed.* **2010**, *49*, 1949–1953.

[178] Matsumoto, Y.; Selim, K. B.; Nakanishi, H.; i. Yamada, K.; Yamamoto, Y.; Tomioka, K. *Tetrahedron Lett.* **2010**, *51*, 404–406.

[179] Hamilton, G. L.; Kang, E. J.; Mba, M.; Toste, F. D. *Science* **2007**, *317*, 496–499.

[180] Hashmi, A. S. K. *Nature* **2007**, *449*, 292–293.

[181] Partyka, D. V.; Robilotto, T. J.; Zeller, M.; Hunter, A. D.; Gray, T. G. *Organometallics* **2008**, *27*, 28–32.

[182] Partyka, D. V.; Robilotto, T. J.; Updegraff, J. B.; Zeller, M.; Hunter, A. D.; Gray, T. G. *Organometallics* **2009**, *28*, 795–801.

[183] Partyka, D. V.; Updegraff, J. B.; Zeller, M.; Hunter, A. D.; Gray, T. G. *Organometallics* **2009**, *28*, 1666–1674.

[184] Brooner, R. E. M.; Widenhoefer, R. A. *Organometallics* **2011**, *30*, 3182–3193.

[185] Brooner, R. E. M.; Widenhoefer, R. A. *Organometallics* **2012**, *31*, 768–771.

[186] Ni, Q.-L.; Jiang, X.-F.; Huang, T.-H.; Wang, X.-J.; Gui, L.-C.; Yang, K.-G. *Organometallics* **2012**, *31*, 2343–2348.

[187] Doherty, S.; Knight, J. G.; Hashmi, A. S. K.; Smyth, C. H.; Ward, N. A. B.; Robson, K. J.; Tweedley, S.; Harrington, R. W.; Clegg, W. *Organometallics* **2010**, *29*, 4139–4147.

[188] Fortman, G. C.; Nolan, S. P. *Organometallics* **2010**, *29*, 4579–4583.

[189] Han, X.; Widenhoefer, R. A. *Angew. Chem. Int. Ed.* **2006**, *45*, 1747–1749.

[190] Hellot, H. Histoire de l' Academie Royale des Sciences. p 101, 1737.

[191] Vogler, A.; Kunkely, H. *Coord. Chem. Rev.* **2001**, *219–221*, 489–507.

[192] Oberbeckmann-Winter, N.; Braunstein, P.; Welter, R. *Organometallics* **2005**, *24*, 3149–3157.

[193] Ito, H.; Saito, T.; Miyahara, T.; Zhong, C.; Sawamura, M. *Organometallics* **2009**, *28*, 4829–4840.

[194] Herrero-Gómez, E.; Nieto-Oberhuber, C.; López, S.; Benet-Buchholz, J.; Echavarren, A. M. *Angew. Chem. Int. Ed.* **2006**, *45*, 5455–5459.

[195] Hintermann, L. Katalytische enantioselektive Fluorierung. Diss. ETH No. 13892, ETH Zurich, 2000.

[196] Assmann, B.; Angermaier, K.; Paul, M.; Riede, J.; Schmidbaur, H. *Ber.* **1995**, *128*, 891–900.

[197] Hong, X.; Cheung, K.-K.; Guo, C.-X.; Che, C.-M. *J. Chem. Soc., Dalton Trans.* **1994**, 1867–1871.

[198] Baenziger, N. C.; Bennett, W. E.; Soborofe, D. M. *Acta Cryst., Sect. B* **1976**, *32*, 962–963.

[199] King, G.; Zheng, S. L.; Coppens, P. CCDC 259382. Priv. Comm. to CCDC, 2004.

[200] Tasjeb, P.; Parkin, A.; Coventry, D.; Parsons, S.; Messenger, D. CCDC 276829. Priv. Comm. to CCDC, 2004.

[201] Jones, P. G.; Maddock, A. G.; Mays, M. J.; Muir, M. M.; Williams, A. F. *J. Chem. Soc., Dalton Trans.* **1977**, 1434–1439.

[202] Wang, M.-Z.; Wong, M.-K.; Che, C.-M. *Chem. Eur. J.* **2008**, *14*, 8353–8364.

[203] Murahashi, S.-I. Ruthenium in Organic Synthesis. Wiley-VCH Verlag GmbH & Co. KGaA, Weinheim, 2005.

[204] Bruneau, C.; Dixneuf, P. H. Topics in Organometallic Chemistry: Ruthenium Catalysts and Fine Chemistry. Springer, Heidelberg, 2011.

[205] Stepnicka, P. Ferrocenes. John Wiley & Sons, Ltd, Chichester, UK, 2008.

[206] Schnyder, A.; Hintermann, L.; Togni, A. *Angew. Chem. Int. Ed. in Engl.* **1995**, *34*, 931–933.

[207] Landert, H.; Spindler, F.; Wyss, A.; Blaser, H.-U.; Pugin, B.; Ribourduoille, Y.; Gschwend, B.; Ramalingam, B.; Pfaltz, A. *Angew. Chem.* **2010**, *122*, 7025–7028.

[208] Pugin, B.; Lotz, M.; Landert, H.; Wyss, A.; Aardoom, R.; Gschwend, B.; Pfaltz, A.; Spindler, F. Bidentate chiral ligands for use in catalytic asymmetric addition reactions. Int. Patent WO 2009/65784 A1, May 28, 2009.

[209] Althaus, M. Asymmetric C–F Bond Formation Catalyzed by Ruthenium PNNP Complexes. Diss. ETH No. 17673, ETH Zurich, 2008.

[210] Eisenberger, P. The Development of New Hypervalent Iodine Reagents for Electrophilic Trifluoromethylation. Diss. ETH No. 17371, ETH Zurich, 2007.

[211] Kieltsch, I. Elektrophile Trifluormethylierung. Diss. ETH No. 17990, ETH Zurich, 2008.

[212] Stanek, K. Electrophilic Trifluoromethylation of Alcohols and Remote Metal-Fluorine Interaction in Late-Transition-Metal Complexes. Diss. ETH No. 18493, ETH Zurich, 2009.

[213] Koller, R. Taking Electrophilic Trifluoromethylation Chemistry a Step Further. Diss. ETH No. 19219, ETH Zurich, 2010.

[214] Niedermann, K. Direct Trifluoromethylation of Organonitrogen Compounds with Hypervalent Iodine Reagents. Diss. ETH No. 20465, ETH Zurich, 2012.

[215] Allen, A. E.; MacMillan, D. W. C. *J. Am. Chem. Soc.* **2010**, *132*, 4986–4987.

[216] Matoušek, V.; Togni, A.; Bizet, V.; Cahard, D. *Org. Lett.* **2011**, *13*, 5762–5765.

[217] Deng, Q.-H.; Wadepohl, H.; Gade, L. H. *J. Am. Chem. Soc.* **2012**, *134*, 10769–10772.

[218] Jensen, S. B.; Rodger, S. J.; Spicer, M. D. *J. Organomet. Chem.* **1998**, *556*, 151–158.

[219] White, C.; Yates, A.; Maitlis, P. M.; Heinekey, D. M. *Inorg. Synth.* **1992**, *29*, 228–234.

[220] Yakelis, N. A.; Bergman, R. G. *Organometallics* **2005**, *24*, 3579–3581.

[221] Brookhart, M.; Grant, B.; Volpe, A. F. *Organometallics* **1992**, *11*, 3920–3922.

[222] Meerwein, H.; Hederich, V.; Wunderlich, K. *Arch. Pharm.* **1958**, *291*, 541–554.

[223] Brandys, M.-C.; Jennings, M. C.; Puddephatt, R. J. *J. Chem. Soc., Dalton Trans.* **2000**, 4601–4606.

[224] Irangu, J. K.; Jordan, R. B. *Inorg. Chem.* **2003**, *42*, 3934–3942.

[225] Bradley, D. C.; Thomas, I. M. *J. Chem. Soc.* **1960**, 3857–3861.

[226] Lin, Y.; Nomiya, K.; Finke, R. G. *Inorg. Chem.* **1993**, *32*, 6040–6045.

[227] Fieser, L. F.; Seligman, A. M. *J. Am. Chem. Soc.* **1939**, *61*, 136–142.

[228] Repine, J. T.; Johnson, D. S.; White, A. D.; Favor, D. A.; Stier, M. A.; Yip, J.; Rankin, T.; Ding, Q.; Maiti, S. N. *Tetrahedron Lett.* **2007**, *48*, 5539–5541.

[229] Marion, N.; Navarro, O.; Mei, J.; Stevens, E. D.; Scott, N. M.; Nolan, S. P. *J. Am. Chem. Soc.* **2006**, *128*, 4101–4111.

[230] Michael, F. E.; Cochran, B. M. *J. Am. Chem. Soc.* **2006**, *128*, 4246–4247.

[231] Schunn, R. A.; Ittel, S. D.; Cushing, M. A. *Inorg. Synth.* **1990**, *28*, 94–98.

[232] Hirabayashi, R.; Ogawa, C.; Sugiura, M.; Kobayashi, S. *J. Am. Chem. Soc.* **2001**, *123*, 9493–9499.

[233] Das, B.; Shinde, D. B.; Kanth, B. S.; Kumar, J. N. *Helv. Chim. Acta* **2011**, *94*, 1477–1480.

[234] SAINT+, Software for CCD Diffractometers, v. 6.01 and SAINT v. 6.02. Bruker AXS Inc., Madison, WI, 2001.

[235] Sheldrick, G. M. *Acta Cryst., Sect. A* **1990**, *46*, 467–473.

[236] Sheldrick, G. M. SHELXL-97 Program for Crystal Structure Refinement. Göttingen, 1999.

[237] Blessing, R. H. *Acta Cryst., Sect. A* **1995**, *51*, 33–38.

[238] Flack, H. D. *Acta Cryst., Sect. A* **1983**, *39*, 876–881.

[239] Bernardinelli, G.; Flack, H. D. *Acta Cryst., Sect. A* **1985**, *41*, 500–511.

[240] *Acta Cryst., Sect. C: Cryst. Struct. Commun.* **2010**, *66*, e1.

[241] Zhang, Z.; Bender, C. F.; Widenhoefer, R. A. *Org. Lett.* **2007**, *9*, 2887–2889.

# i want morebooks!

Buy your books fast and straightforward online - at one of world's fastest growing online book stores! Environmentally sound due to Print-on-Demand technologies.

## Buy your books online at
## www.get-morebooks.com

Kaufen Sie Ihre Bücher schnell und unkompliziert online – auf einer der am schnellsten wachsenden Buchhandelsplattformen weltweit! Dank Print-On-Demand umwelt- und ressourcenschonend produziert.

## Bücher schneller online kaufen
## www.morebooks.de

VDM Verlagsservicegesellschaft mbH
Heinrich-Böcking-Str. 6-8
D - 66121 Saarbrücken

Telefon: +49 681 3720 174
Telefax: +49 681 3720 1749

info@vdm-vsg.de
www.vdm-vsg.de

Printed by Books on Demand GmbH, Norderstedt / Germany